Cooperative Learning
Making Connections in General Biology

Mimi Bres
Prince George's Community College

Arnold Weisshaar
Prince George's Community College

Australia • Brazil • Japan • Korea • Mexico • Singapore • Spain • United Kingdom • United States

Cooperative Learning: Making Connections in General Biology

Mimi Bres, Arnold Weisshaar

Publisher: Jack Carey

Project Development Editor: Kristin Milotich

Marketing Team: Tami Cueny and Kelly Fielding

Editorial Assistant: Susan Lussier

Production Coordinator: Dorothy Bell

Cover Design: Vernon T. Boes

For product information and technology assistance, contact us at
Cengage Learning Customer & Sales Support, 1-800-354-9706

For permission to use material from this text or product,
submit all requests online at **cengage.com/permissions**
Further permissions questions can be emailed to
permissionrequest@cengage.com

ISBN-13: 978-0-534-37605-5

ISBN-10: 0-534-37605-3

Brooks/Cole
10 Davis Drive
Belmont, CA 94002-3098
USA

Cengage Learning is a leading provider of customized learning solutions with office locations around the globe, including Singapore, the United Kingdom, Australia, Mexico, Brazil, and Japan. Locate your local office at: **international.cengage.com/region**

Cengage Learning products are represented in Canada by Nelson Education, Ltd.

For your course and learning solutions, visit **academic.cengage.com**

Purchase any of our products at your local college store or at our preferred online store **www.ichapters.com**

Printed in the United States of America
7 8 9 10 11 12 11 10 09 08

CONTENTS OVERVIEW

CONTENTS

V. Enzymes and Metabolism

VI. Mitosis and Meiosis

VII. Patterns of Inheritance

VIII. More Patterns of Inheritance

IX. Molecular Genetics

PREFACE: TO THE STUDENT

Cooperative Learning: Making Connections in General Biology has been designed as a supplement to help you get the most out of your introductory biology course. Doing these cooperative activities will provide you with an opportunity to approach lecture concepts in a new way. They will help you:

- reinforce basic concepts covered during lecture
- make connections between the chapters and topics presented in your textbook
- develop graphing and problem solving skills
- analyze data and draw conclusions from experimental results and master terminology.

The skills of science can greatly benefit your life. Science and good thinking go hand in hand. The scientific method can be viewed as a way of using critical and creative thinking to learn about the world. Developing your skills in this course will also enhance your success in other courses.

Practice makes perfect! The more different ways you approach a concept, the more likely you are to be able to recognize and recall this information under stressful conditions (such as an exam day). Work through these exercises during class time, but also review them during your study time. Take advantage of the opportunity to interact with your peers. Studying in groups will often lead to a significant improvement in your exam grades.

Through the use of cooperative activities, you will gain confidence and expertise in logical thinking, the ability to work with others, analyzing what you learn, and communicating your ideas to others. All of these are skills highly sought after in today's competitive employment marketplace.

We hope that working through these cooperative activities will be a rewarding experience, and one that will enhance your success in this course and in your future endeavors.

Mimi Bres and Arnold Weisshaar
Professors of Biology
Prince George's Community College
Largo, MD

ACTIVITY 1-1: TESTING HYPOTHESES WITH EXPERIMENTS

(comprehension)

Anthrax is a fatal, contagious disease that affects farm animals and humans. This disease has received a lot of media attention because it's one of several that can be used as biological warfare agents. To learn more about this disease, scientists at the Biological Warfare Research Center at Fort Dietrich, Maryland set up the following experiment:

One herd of 100 cows was injected with liquid containing anthrax bacteria that had been heated enough to weaken, but not kill the bacteria. A second herd of 100 cows was injected with the same liquid, minus the bacteria. A few weeks later, the scientists injected both herds of 100 cows with full strength anthrax bacteria. The cattle that had been previously injected with weakened bacteria showed no symptoms of the disease. The cattle from the second herd, however, became ill with anthrax and died.

1. What hypothesis was being tested in this experiment?

2. Which herd of cattle represents the control group? ______________________________

3. Why was a control group needed in this experiment?

4. Why were both groups of cattle injected, even though one group didn't receive any bacteria in the injection?

5. A Senator from the Armed Forces Appropriations Committee paid a visit to Fort Dietrick to evaluate the experiments. The Senator was critical of the large number of animals being used and suggested that it would be cheaper to use only 10 animals in each group. How would you respond to the Senator?

ACTIVITY 1-2: ANALYZING THE RESULTS OF AN EXPERIMENT

(comprehension)

A group of ecology students was investigating the food chain in a nearby stream last summer, and noticed brook trout feeding on mayfly nymphs among the rocks on the stream bottom. Mayfly nymphs are large crawling insects that form one stage in the development of mayflies. The students hypothesized that the trout were reducing the insect population by at least 30%. To test this hypothesis, the students set up a series of wire mesh cages on the floor of the stream. The cages allowed the mayfly nymphs to enter and leave as they pleased, but prevented access by the trout.

They planned to compare the number of insects found under the cages with the number of insects in a nearby section of the stream that did not have cages to keep the fish away. Each week, for five weeks, the students collected and counted all the insects from the protected and unprotected areas of the stream. **Table 1-1** shows the results of their experiment:

Table 1-1. Results of Predation Experiment

Number of Nymphs Collected	Caged Areas	Uncaged Areas
Week 1	66	50
Week 2	120	61
Week 3	141	74
Week 4	176	82
Week 5	234	93
Total Nymphs Collected	737	360
Average Number of Nymphs Collected Over the Five Week Period	147.4	72.0

1. Which was the control group in this experiment?

   a. the caged areas
   b. the uncaged areas
   c. no control group was used in this experiment

2. **Using the graph paper in Figure 1-1,** graph the results of this experiment. **Make sure to label the horizontal and vertical axes and give your graph a title.**

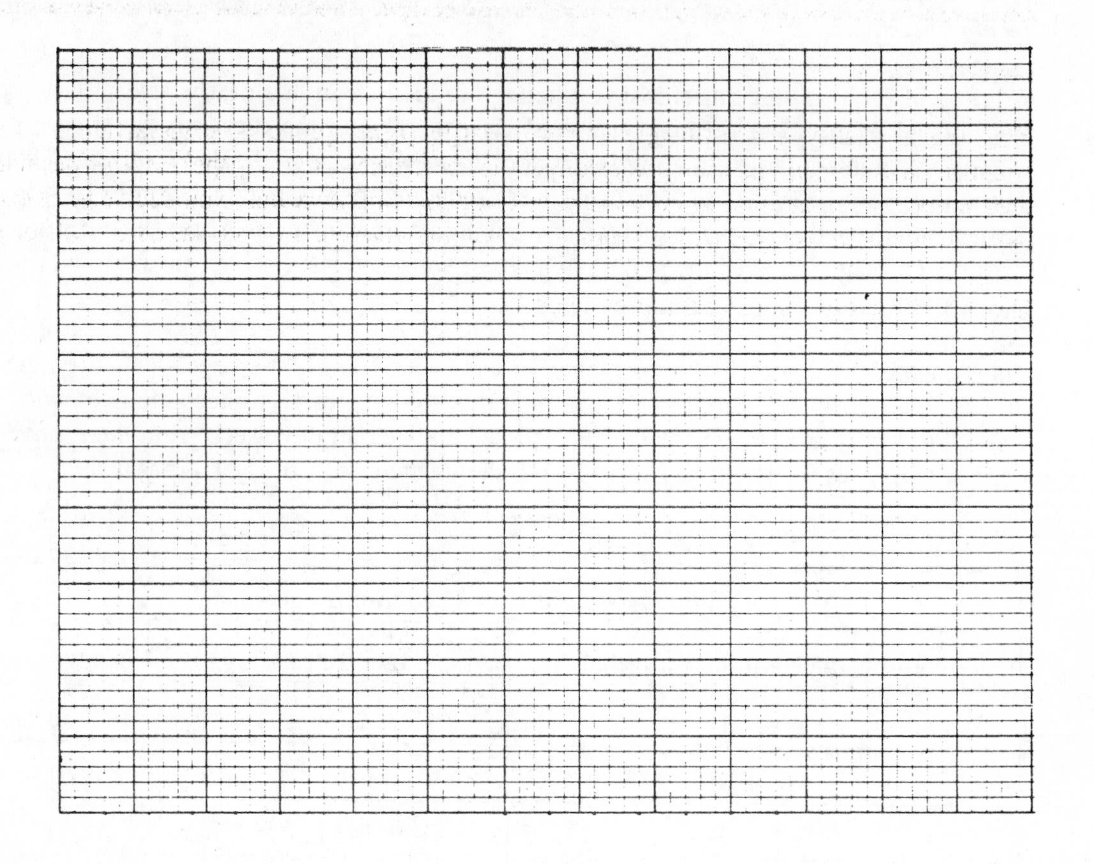

Figure 1-1.

3. From Week 1 through Week 5, the number of nymphs collected in both the caged and uncaged areas:

   a. increased b. decreased c. remained the same

4. Do the results of the experiment support the hypothesis? **Explain your answer.**

ACTIVITY 1-3: A COMMERCIAL APPLICATION FOR THE SCIENTIFIC METHOD

(application)

You are a quality control engineer for a large paper products manufacturer. Several commercial customers have complained that the absorbency of your paper towels does not meet their needs. In an attempt to improve your company's product line, you decide to conduct an experiment to see how your towels compare to your competitor's towels in terms of absorbency.

Describe your experimental setup in detail.

ACTIVITY 1-4: UNDERSTANDING THE SCIENTIFIC METHOD

(application)

Student A carried out an experiment to test for the presence of simple sugar in a potato. He did the experiment using a chemical indicator called Benedict's solution, which turns bright orange when heated in the presence of simple sugars. He had previously verified that the indicator solution was effective by testing it on a known simple sugar, glucose.

To carry out his potato experiment, he added the following to his experimental test tube:

ground up potato
distilled water
10 drops of glucose solution
5 ml of Benedict's solution

Following a few minutes of heating, the Benedict's solution turned bright orange. The student concluded that potato contains simple sugars.

1. Do you agree with his conclusions? **Explain your answer**.

2. What would be an appropriate **control** for this experiment?

ACTIVITY 2-1: ATOMIC STRUCTURE

(recall and comprehension)

1. Draw a diagram of a phosphorus atom (atomic number = 15), showing the **nucleus** and number of **electrons** in each shell.

2. Draw a dot model of a **chlorine** atom (atomic number = 17), showing only the number of electrons in the **outer** shell.

   Example:
   (aluminum, atomic number = 13)

   •

   • **Al** •

   Dot Model

3. Explain how it is possible for **one** carbon atom (atomic number = 6) to bond to **four** hydrogen atoms (atomic number = 1). Include a **diagram** showing this bonding pattern.

ACTIVITY 2-2: ATOMS AND BONDS

(recall and comprehension)

Test your knowledge. Match the statements with the best answer below. Answers can be used **more than once.**

A.	**IONIC BOND**	**F.**	**NUMBER OF PROTONS**
B.	**SINGLE COVALENT BOND**	**G.**	**OUTER ELECTRON SHELL**
C.	**DOUBLE COVALENT BOND**	**H.**	**WATER**
D.	**POLAR COVALENT BOND**	**I.**	**ELECTRON**
E.	**NITROGEN**	**J.**	**NONPOLAR COVALENT BOND**

1. _____ formed by the sharing of **two** pairs of electrons

2. _____ determines the atomic number of an element

3. _____ formed when atoms share a pair of electrons from their outer shells, but the electrons are **not** shared equally

4. _____ an example of a molecule with a **polar** covalent bond

5. _____ has an equal number of protons and electrons

6. _____ type of bond found in a molecule that **cannot** dissolve in water

7. _____ formed when two atoms share pairs of electrons from their outer shells, and the electrons are shared **equally**

8. _____ location of electrons that determine how an atom will form a chemical bond

9. _____ bond formed when two atoms share **one pair** of electrons from their outer shells

10. _____ negatively charged particles of an atom

11. _____ type of bond found in a water molecule

12. _____ type of bond commonly found in **water soluble** organic molecules

ACTIVITY 2-3: PROPERTIES OF WATER

(comprehension)

The following paragraph contains many errors. Use the space provided between the sentences to correct the mistakes as necessary, ensuring that the entire paragraph is error-free.

Water is an element that is essential for life. This unique atom displays many

characteristics that account for its vital role on earth. Water's features are a result of

the chemical bonding of hydrogen and oxygen. Hydrogen has a higher affinity for

protons than does oxygen. This results in partial negative and partial positive charges

on the water molecules. Water molecules, therefore, are nonpolar. Since opposite

charges attract, an ionic bond is formed between adjacent water molecules. Water

molecules are also attracted to other nonpolar molecules in the same manner, a

characteristic called adhesion. Water also has a low specific heat (defined as the energy

required to change water from a liquid to a gas). This physical characteristic is

important in the maintenance of body temperature. As water evaporates from our

bodies, we are able to absorb energy, cooling the body. In addition, the nonpolar

nature of water allows it to serve as an almost universal solute.

COOPERATIVE ACTIVITY 2-4: SYNTHESIS AND HYDROLYSIS

(comprehension)

Answer the following questions using these vocabulary words: **monomer, hydrolysis, dehydration synthesis (or condensation), water, and polymer.** All the questions refer to the diagram in **Figure 2-1**.

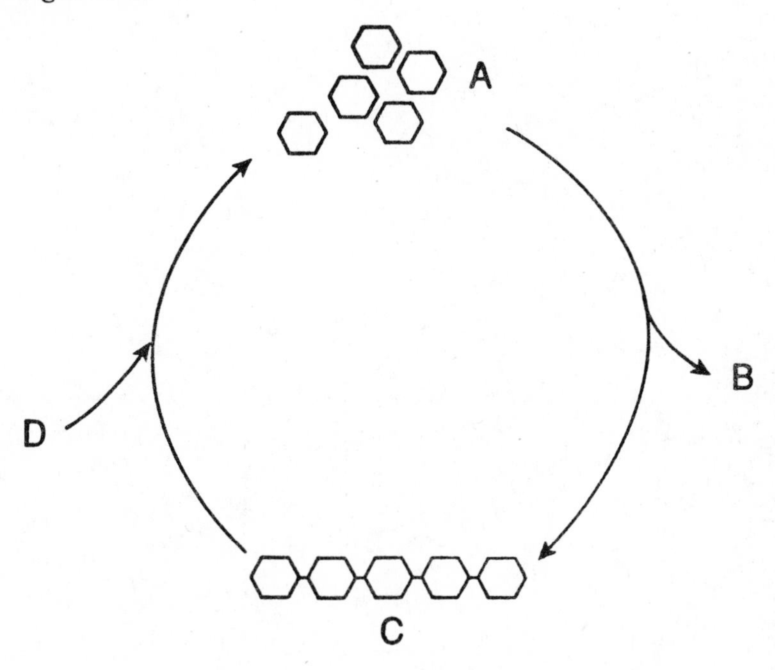

Figure 2-1. Synthesis and Hydrolysis

1. What process is taking place along the line that begins with the letter A and ends with the letter C? ___________

2. Name the molecule(s) marked with the letter A _______________

3. Name the molecule(s) marked with the letter C _______________

4. What process is taking place along the line that begins with the letter C and ends with the letter A? ___________

5. Name the molecule(s) being inserted at D _______________

6. Name the molecule(s) being removed at B _______________

7. How **MANY** of the molecules from question #6 were removed to make the molecule marked with the letter C? ___________

ACTIVITY 2-5: WATER AT WORK

(comprehension)

Do the special **properties of water** play an important role in any of the following activities? If so, explain how. If not, explain why water is not important for that function.

1. Homeostatic control of body temperature -

2. Hydrolysis of polysaccharides (such as starch) -

3. Survival of fish in a lake covered with ice -

4. Uptake of oxygen in the alveolus of an athlete during vigorous exercise -

ACTIVITY 2-6: BUILDING CARBOHYDRATES

(application)

Look at the diagram in **Figure 2-2** and answer the following questions:

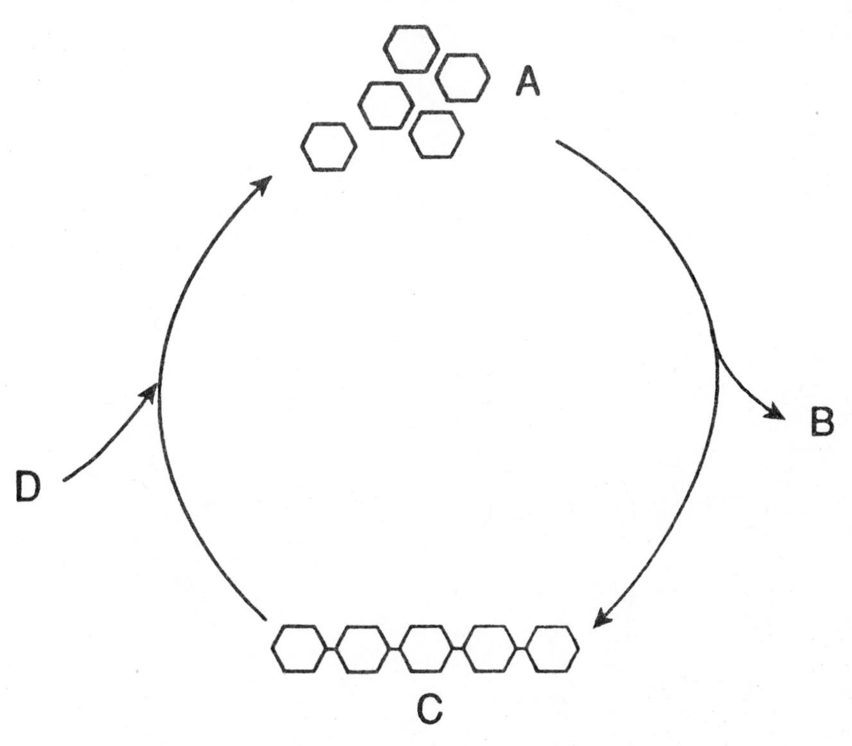

Figure 2-2. Synthesis of Carbohydrates

1. If glycogen were hydrolyzed, which letter would represent the result? ______

2. Which letter represents monosaccharides? ______

3. If this diagram represents the hydrolysis of glycogen, name the monosaccharides produced ______________________________

4. If you were to attach all the monosaccharides shown in this diagram to the **existing** glycogen molecule, how many **additional** covalent bonds would be needed? ______

ACTIVITY 2-7: COMPLEX CARBOHYDRATES

(comprehension)

1. Saliva contains an enzyme that digests (breaks down) starch. If you were to chew a piece of bread for a few minutes **WITHOUT SWALLOWING**, you would notice a change in the way it tastes. At first it will taste as you expect bread to taste, but after a few minutes, the bread begins to taste sweet. Using your knowledge of complex carbohydrates, explain this unexpected result.

2. **Challenge Question!**

   Would you expect the same result if you were eating **celery**? **Explain your answer**.

ACTIVITY 2-8: REVIEW OF ORGANIC COMPOUNDS

(recall)

Organic Compound	Main Function
glucose	
	polysaccharide used for energy storage
estrogen, progesterone, testosterone	
fats	
	protein that provides strength to skin, hair, and nails
	regulates chemical reactions in the body
hemoglobin	
insulin	
	protein that helps with immune system defense
	molecule that releases a form of energy that can be used directly by cells
	structural material in plant cell wall
phospholipids	
DNA	

ACTIVITY 3-1: REVIEW OF CELLULAR ORGANELLES

(recall)

Choose the most appropriate answer for the questions below. Answers can be used **more than once**.

A.	**NUCLEUS**	**E.**	**CHLOROPLASTS**	**I.**	**LYSOSOMES**
B.	**CELL MEMBRANE**	**F.**	**CELL WALL**	**J.**	**CYTOPLASM**
C.	**RIBOSOMES**	**G.**	**MITOCHONDRIA**	**K.**	**GOLGI BODIES**
D.	**CILIA AND/OR FLAGELLA**	**H.**	**ROUGH ER**	**L.**	**SMOOTH ER**

1. H ___ location where collagen and melanin for the skin are manufactured
2. K ___ packages manufactured chemicals for transport out of the cell
3. G ___ are most active during vigorous exercise
4. I ___ are used by white blood cells to "digest" and destroy bacteria
5. L ___ manufactures sex hormones
6. C ___ synthesizes more material for newly formed muscle cells than for newly formed fat cells
7. J ___ liquid or jelly-like material inside the cell
8. F ___ made of cellulose
9. B ___ regulates movement of materials in and out of the cell
10. A ___ organelle missing from mature human red blood cells
11. A ___ contains the chromosomes
12. E ___ organelle responsible for the bumper sticker, "Have you thanked a green plant today?"
13. D ___ important for locomotion of one-celled organisms

ACTIVITY 3-2: ORGANELLE FUNCTION

(application)

1. In a recent murder mystery, a woman died in minutes after consuming cyanide-laced sugar. Forensic scientists did some research and found out that death was caused by lack of the energy rich molecule ATP. Which cellular organelle was affected by this poison? **Explain your answer.**

   Mitochondria, because most ATP is made here.

2. Examination of a sample of glandular cells from an unknown body location reveals an extensive network of smooth endoplasmic reticulum. Circle **all** substances from the following list that might be produced by these cells. **Explain your answers.**

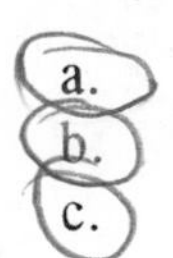

a.	skin oils	d.	cholesterol
b.	digestive enzymes	e.	ear wax
c.	sex hormones	f.	cellulose

3. A person who smokes destroys the cilia on the columnar epithelial cells that line the trachea. In most cases, the cilia will regenerate if a person quits smoking. If the cilia did not regenerate or did not resume proper function, what cellular organelle might be implicated in the problem? **Explain your answer.**

   Micro t

4. In a classroom experiment, you ground up some spinach in distilled water and filtered the liquid onto a piece of filter paper. The isolated organelles absorbed carbon dioxide and released oxygen. Which cellular organelle was most likely on the filter paper? **Explain your answer**.

Chloropast,

Challenge Question!!

5. In reference to question #4, would you expect the same results if you ground up some mushrooms? **Explain your answer**.

No, because they are not plants.

ACTIVITY 3-4: PREDICTING MOLECULAR MOVEMENT

(comprehension)

1	2	3	4	5	6	7	8	9

1. You prepare a plate of agar (a gelatin-like substance) as shown in the figure above. At position 9, you place iodine solution, an indicator chemical that turns black in the presence of starch. At position 1, you place starch solution.

   At what position in the agar plate do you predict the black color will first appear? **Explain your answer.**

2. In reference to question 1, is the chemical reaction that produced the black color the result of molecules moving by diffusion or osmosis? **Explain your answer.**

3. If you repeated the experiment, placing an amino acid solution at one end of the gel and a protein solution at the other end, which solution would reach position 5 first? **Explain your answer.**

Extension!

4. In the laboratory, set up experiments to verify the hypotheses you made concerning the movement of various substances (such as starch, iodine, amino acids, and proteins) through an agar gel.

ACTIVITY 3-5: EXPERIMENTS WITH DIFFUSION AND OSMOSIS

(application)

A student poured 250 ml of starch solution into each of two 500 ml containers. Container A was filled with a 40% starch solution, and container B with a 20% starch solution. After weighing, the two containers were connected with a selectively permeable membrane that was **permeable to small molecules such as water or glucose, but not to larger molecules such as starch.** After 30 minutes, the beakers were separated and weighed again.

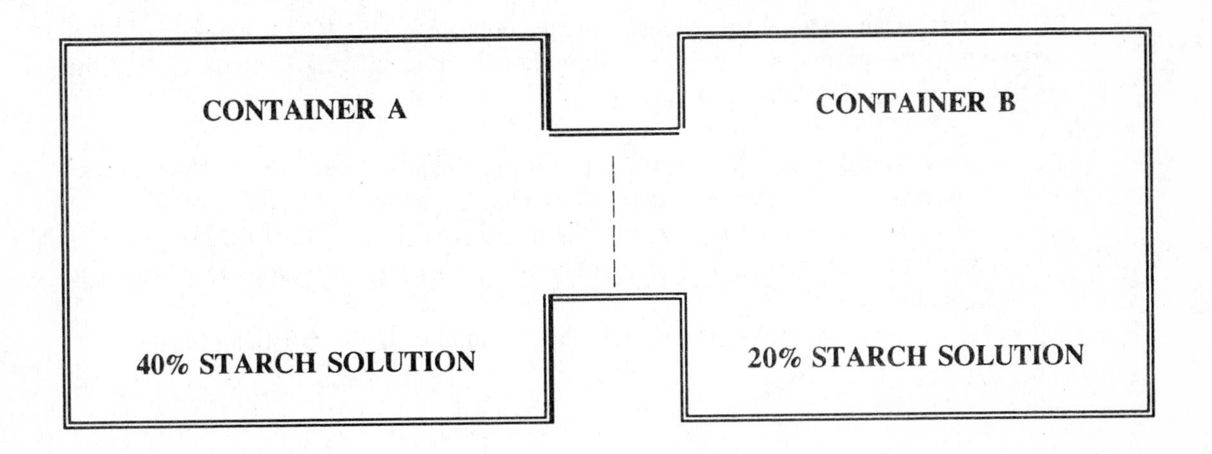

Figure 3-1. Starch Experiment

1. When the experiment began, container A was __________ to container B.

   a. hypertonic
   b. isotonic
   c. hypotonic

2. Will the starch molecules move from one container to the other? If so, why? If not, why not?

3. Will water molecules move from one container to the other? **Explain your answer.**

4. Will diffusion and/or osmosis occur between the two connected containers? **Explain your answer**.

Challenge Questions!!

5. Saliva contains an enzyme that digests (breaks down) starch. If you were to add this enzyme to both beakers, which of the following would be most likely to occur over the next 60 minutes? **Explain your answer**.

   a. there would be no change in the pattern of diffusion between the two beakers
   b. there would be a net movement of glucose from Beaker A to Beaker B
   c. there would be a net movement of glucose from Beaker B to Beaker A
   d. the selectively permeable membrane will prevent the breakdown of starch to glucose
   e. glucose cannot move from one beaker to another if starch is also present

6. After two hours, you continued the experiment by emptying the contents of Beaker A and Beaker B into separate graduated cylinders. Would the volume of liquid poured out of each beaker still equal 250 ml? **Explain your answer**.

ACTIVITY 3-6: DIFFUSION AND OSMOSIS AFFECT OUR EVERYDAY LIVES

(comprehension)

Using your knowledge of the principles of diffusion and osmosis, **explain the reason** for the following:

1. In some restaurants, sliced potatoes are soaked in water before they are fried.

2. You are reading the label of several brands of eye drops and notice that the contents of all the brands are isotonic to body fluids.

3. You are planning to fertilize your lawn. You think to yourself, "If I used more fertilizer, I wouldn't have to do this as often." You apply the fertilizer in double the concentration recommended on the label. The grass turns brown and dies.

4. A student performed the following experiment. She added salt to the water in a pot containing tomato plants. After 15 minutes, she observed wilting begin. After 45 minutes wilting was pronounced and the plant was collapsing. After 90 minutes, the plant was severely wilted.

ACTIVITY 3-7: REVIEW OF MEMBRANE TRANSPORT

(recall)

For each of the following molecules, write **D** if it is moved across cell membranes by **simple diffusion**, **P** if it is moved by **passive** transport (such as facilitated diffusion), and **A** if it requires **active** transport.

1. ____ amino acids
2. ____ Na^+
3. ____ CO_2
4. ____ K^+
5. ____ cholesterol
6. ____ glucose
7. ____ H_2O
8. ____ Cl^-
9. ____ O_2
10. ____ estrogen or testosterone

11. Consider a growing clam in the ocean that needs calcium ions to increase the size of its shell. Since the calcium concentration is higher inside the shell-secreting cells than in the ocean, how could the cells obtain more calcium? **Be specific.**

12. If molecules enter a cell by **endocytosis**, have they passed through the cell membrane? **Explain your answer**.

13. When a vesicle enters a cell by endocytosis, how are the contents of that vesicle released into the cytoplasm? **Be specific**.

ACTIVITY 3-8: PRACTICAL APPLICATIONS OF MEMBRANE TRANSPORT

(application)

There is a direct relationship between the effectiveness of a general anaesthetic (such as ether, chloroform, or nitrous oxide) and its lipid solubility. Local anesthetics (such as procaine and lidocaine) and systemic depressants (such as barbiturates and alcohol) are also lipid soluble.

1. Why would lipid solubility enhance the speed and effectiveness of the above chemicals?

2. If a drug had the ability to block ion channels or inhibit ion exchange pumps (such as the sodium-potassium pump), would it have anaesthetic properties? **Explain your answer.**

ACTIVITY 4-1: ASTEROIDS AND DINOSAURS

(comprehension)

One hypothesis concerning the extinction of the dinosaurs 65 million years ago states that a giant asteroid collided with the planet Earth. As a result of this collision, a huge cloud of dust filled the atmosphere. Sunlight would have been blocked for several months or even longer.

1. How might this affect plants around the world? **Explain your answer**.

2. Suggest **two** ways in which a problem with plant life could affect dinosaur populations. **Be specific. Explain each answer.**

   a)

   b)

ACTIVITY 4-2: BRIGHT BEANS

(comprehension)

To demonstrate plant growth, a student set up a bean plant "maze." He cut holes in the partitions of a divided corrugated carton to form a maze (see **Figure 4-1**). At one end of the carton, he cut an opening to allow sunlight to enter. A pot containing a small bean plant was set in the corner of the carton that was furthest from the outside opening.

The carton was tightly covered and sealed with tape so that no light could enter the box through the lid. The box was placed on the window sill with the outside opening facing the light. After several weeks, the student noticed that the bean plant had negotiated the maze and was growing out of the opening.

1. Suggest a probable reason for the observed growth pattern.

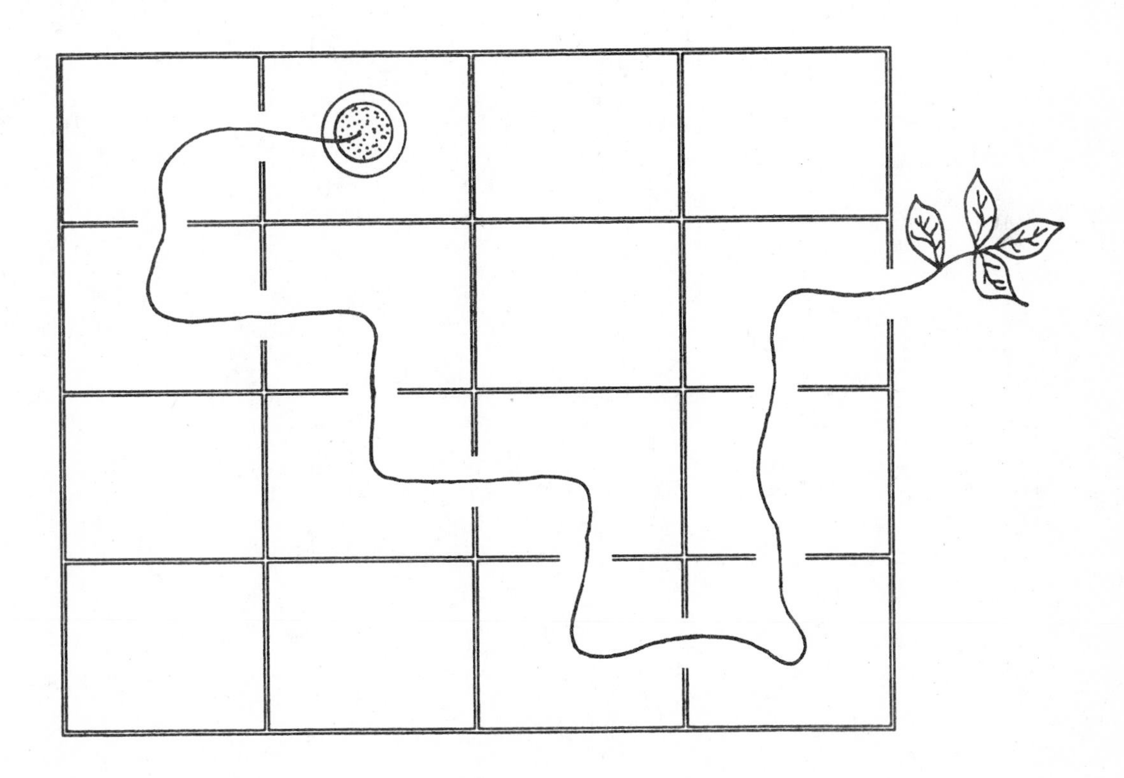

Figure 4-1. Bean Plant Maze

ACTIVITY 4-3: POND WATER, PLANTS, AND SNAILS

(application)

A classmate performed the following experiment. She compared five containers, whose contents are listed below:

Container 1	Pond water only
Container 2	Pond water and a snail
Container 3	Pond water and two snails
Container 4	Pond water and a small aquatic plant
Container 5	Pond water, a small aquatic plant, and a snail

1. Is there a control in this experiment? If so, which container is the control?

2. What hypothesis was the student probably testing in this experiment? There are several possibilities. Can you think of more than one?

3. **Not considering the issue of food supply**, in which of the above containers would the snail live longest?

4. **Again, do not take food supply into account.** If a goldfish was added to each container, in which container would it live the longest? **Explain your answer**.

5. How would the results of the experiment change if the containers were kept in the **dark**?

6. Which container has the best balance between photosynthesis and cell respiration? **Explain your answer**.

ACTIVITY 4-4: CELLS AND ENERGY

(recall and comprehension)

A) $C_6 H_{12} O_6 + 6 O_2$ -----------> $6 CO_2 + 6 H_2O + ATP$

B) $6 CO_2 + 12 H_2O$ -----------> $C_6 H_{12} O_6 + 6 H_2O + 6 O_2$

1. Name the process described by **equation A**: Photosynthesis

2. Name the process described by **equation B**: Cell respiration

3. **Equation A** takes place inside which cellular organelle? ~~Cytoplasm~~ Mito

4. In plant cells, **equation B** takes place inside which cellular organelle?

   ~~Mitochondria~~ chloroplast

5. What **two items NOT** listed in **equation B** are required for this process to take place?

   Sunlight **and** Chloropyll

6. From the list below, circle the names of all organisms that can perform **both equation A and equation B.**

human being	grass	sparrow
earthworm	mushroom	cherry tree
phytoplankton	bread mold	corn

7. Which equation takes in **inorganic** molecules and releases **organic** molecules? B

8. Which equation would contribute more to the problem of global warming (the greenhouse effect)? **Explain your answer**.

A, add CO2 to atmosphere

9. Since plants don't move very much, do they need ATP energy? **Explain your answer**.

Yes, need to complete bio-process

Challenge Question!!

10. Would the answer you listed for question #4 still be appropriate for photosynthetic bacteria? **Explain your answer**.

No, dont have organelles

ACTIVITY 4-5: COMPARING PHOTOSYNTHESIS WITH CELL RESPIRATION

(recall and comprehension)

Decide between yes or no for each item below.

Key Points	Cell Respiration	Photosynthesis
requires molecules of carbon dioxide gas to start	YES NO	YES NO
requires glucose or other organic (food) molecules to start	YES NO	YES NO
requires molecules of oxygen to complete	YES NO	YES NO
requires water molecules to start	YES NO	YES NO
requires light energy to start	YES NO	YES NO
produces glucose molecules	YES NO	YES NO
produces molecules of carbon dioxide	YES NO	YES NO
produces molecules of oxygen	YES NO	YES NO
produces ATP energy from food energy	YES NO	YES NO
humans and other animals do this	YES NO	YES NO
decomposers and detritus feeders do this	YES NO	YES NO
algae and other green plants do this	YES NO	YES NO

ACTIVITY 5-1: EFFECT OF pH ON ENZYME ACTIVITY

(comprehension)

Table 5-1 contains the results of an experiment that examined the effect of pH on the activity of an enzyme isolated from the **human digestive tract**. The **+/- signs** indicate the presence or absence of enzyme activity.

Table 5-1. Presence of Absence of Enzyme Activity at Varying pH Levels

Tube	pH	Elapsed Time (Minutes)											
		2	4	6	8	10	12	14	16	18	20	22	24
1	1.1	-	-	-	-	-	-	-	-	-	-	-	-
2	1.3	-	-	-	-	-	-	-	-	-	-	-	-
3	1.5	-	-	-	+	+	+	+	+	+	+	+	+
4	1.7	+	+	+	+	+	+	+	+	+	+	+	+
5	1.9	-	-	-	-	-	-	+	+	+	+	+	+
6	2.1	-	-	-	-	-	-	-	-	-	-	-	-

1. Summarize the **results** of this experiment.

2. Would you characterize this enzyme as being **general or specific** in regard to its **pH requirements? Explain your answer**.

3. What conclusions can you draw about this enzyme and its function?

Challenge Questions!!

4. Assume you took two antacid tablets after eating. How would this affect the activity of the enzyme mentioned in **Table 5-1**?

5. Which **test tube number** would most closely resemble the expected results of the situation mention in question #4?

ACTIVITY 5-2: EFFECT OF SUBSTRATE CONCENTRATION ON ENZYME ACTIVITY

(application)

Examine the graph in **Figure 5-1** in which the **rate** of an enzyme catalyzed reaction is plotted for different concentration of **substrate (S)**.

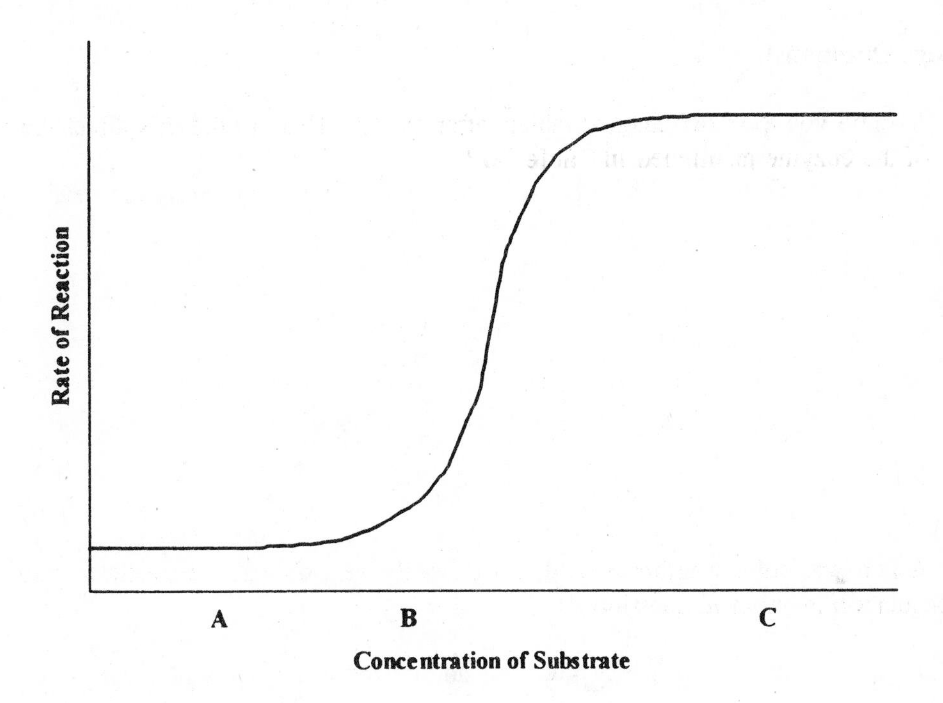

Figure 5-1. Effect of Substrate Concentration on Reaction Rate

In reference to the graph, **explain** why **each** of the following statements is **true or false**.

1. If the concentration of S is at **point C** on the graph, then addition of more S will have very little effect on the reaction rate.

2. If the concentration of S is at **point C**, addition of more enzyme will increase the reaction rate.

3. If the concentration of S is at **point B**, addition of S will speed up the reaction.

4. If the concentration of S is at **point A**, adding more enzyme will greatly increase the reaction rate.

ACTIVITY 5-3: ARE ENZYMES UNIVERSAL?

(comprehension)

Do one-celled organisms such as protists have enzymes? **If so, explain your answer and give some examples** of enzyme-catalyzed reactions that could occur in these organisms (names of specific enzymes are not necessary). If protists **don't** have enzymes, just **explain your answer**.

ACTIVITY 5-4: NEW USES FOR ENZYMES

(application)

A genetically engineered form of corn produces a chemical that is toxic to corn borers (an insect pest that destroys 13 billion ears of corn each year). Synthesis of the chemical toxin requires several steps, each catalyzed by a different enzyme (as shown below):

	enzyme 1		**enzyme 2**	
A	**--------------->**	**B**	**--------------->**	**corn borer toxin**

Compound A is a nutrient absorbed from the soil. Within the plant, compound A is converted into compound B through the action of enzyme 1. In the second step of the synthesis, compound B is converted to the corn-borer toxin through the action of enzyme 2.

After several successful seasons of corn growing, the researchers discovered a few mutant corn plants that could only produce the corn-borer toxin when compound B itself was added to the soil.

1. Which of the following would be valid **conclusions** based on the above information:

   a. the mutation altered the genetic information needed to produce enzyme 2
   b. the mutants are producing excessive amounts of enzyme 2
   c. the mutation altered the genetic information needed to produce enzyme 1
   d. the mutation altered the genetic information needed to produce both enzyme 1 and enzyme 2
   e. the mutation hasn't affected the production of either enzyme 1 or enzyme 2, but it has affected the ability of the plant to absorb chemicals from the soil

   Explain why EACH of the above conclusions is valid or invalid.

2. If an extra amount of **compound A** were added to the soil, would it improve the ability of the mutants to produce the corn-borer toxin? **Explain your answer**.

ACTIVITY 6-1: IDENTIFYING THE STAGES OF MITOSIS

(recall)

Identify the following stages of the cell cycle and mitotic cell division:

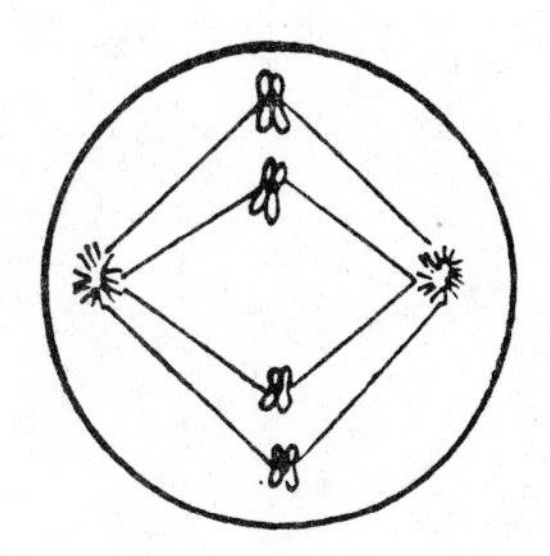

1. Metaphase

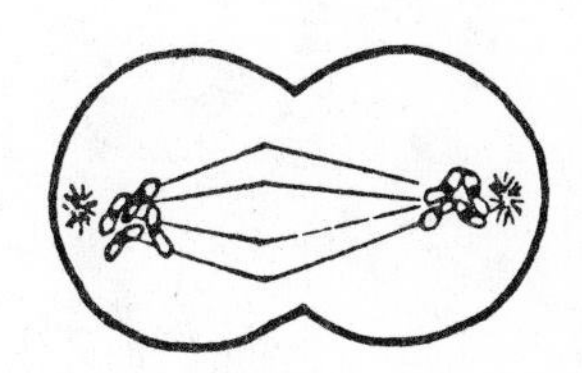

2. Anaphase

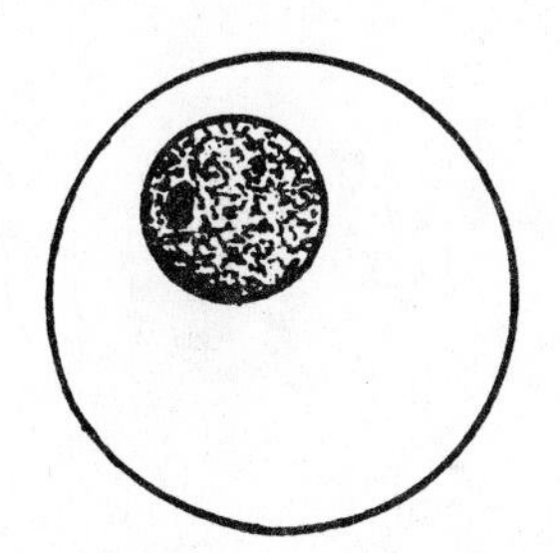

3. ~~Cytokinesis~~

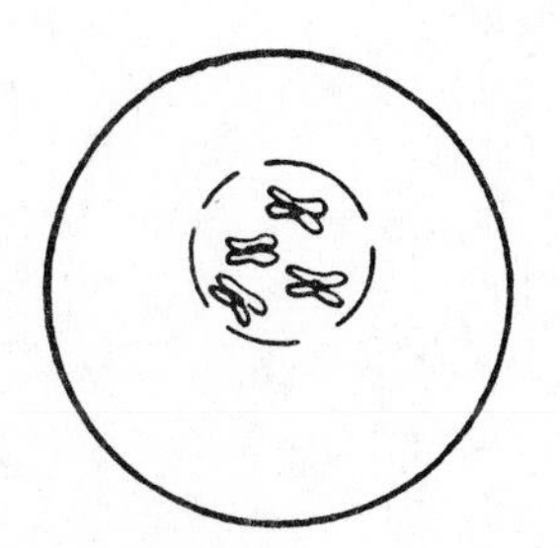

4. ____________

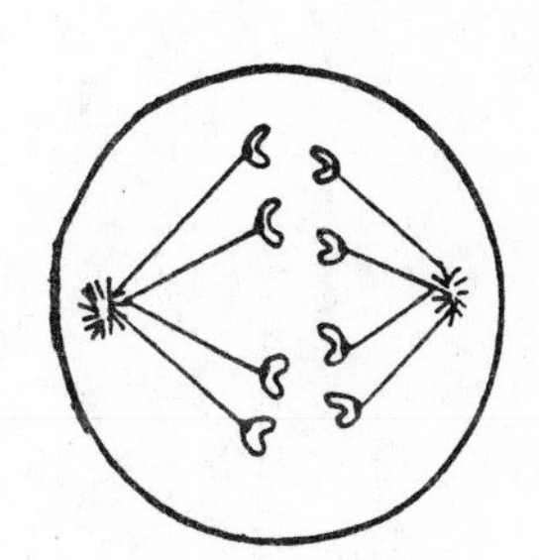

5. ____________

ACTIVITY 6-2: MITOSIS AND CELL REPLACEMENT

(comprehension)

Each minute, 300 million of your body cells die! If these cells were not replaced, all would be dead in only 230 days. Not all body cells age and die at the same rate, however. Some cells last for years, while others are replaced on a daily basis.

List **three types of body cells** that must be replaced **frequently**. **Explain your answer**.

Skin cells

blood cells

Sperm cells & egg cells

alinging of stomach.

ACTIVITY 6-3: STAGES OF THE CELL CYCLE IN MEIOSIS

(recall)

Complete the table by entering the appropriate stage of the cell cycle or an important activity that occurs during that stage.

Stages of the Cell Cycle	Cellular Activity
Meiosis I Prophase I	crossing over occurs
anaphase II	sister move apart
Anaphase I	members of homologous pairs separate from each other
telophase I	
interphase	replicate DNA
	pairs of homologous chromosomes located at the equator
telophase II	4 daughter cells created
telophase I	first cytokinesis occurs
prophase II	tetrads form
metaphase II	chromosomes align

ACTIVITY 6-6: COMPARING MITOSIS AND MEIOSIS

(recall)

Complete this chart with **yes or no** to distinguish mitosis from meiosis.

Key Points	Mitosis	Meiosis
produces two diploid daughter cells	Yes	
produces four haploid daughter cells		Yes
completed after one division	Yes	
crossing over occurs between homologous chromosomes		Yes
produces gametes		Yes
allows for growth and cell replacement	Yes	
involves duplication of chromosomes	Yes	Yes
tetrads form		Yes
cheek cells are produced by this process	Yes	
occurs during budding or fission	Yes	
occurs in sperm and egg formation		Yes
occurs during embryonic development, beginning right after fertilization	Yes	
occurs during production of gametes in the ovaries or testes		Yes
occurs when the immune system produces armies of defensive white blood cells	Yes	
can be used to explain why three brothers do not look alike		Yes
occurs when a seed sprouts and grows into an adult plant	Yes	
occurs during the healing process after an injury	Yes	

ACTIVITY 7-1: GENES AND ALLELES

(recall and comprehension)

This diagram represents the genes along one section of a pair of homologous chromosomes (remember that chromosomes may contain thousands of genes). If the alleles on both chromosomes are **alike**, label them **homozygous**. If the two alleles are **different**, label them **heterozygous**.

1. hetero
2. homo
3. homo
4. homo
5. hetro

B	b
h	h
D	D
A	A
T	t

6. How many traits are represented on these two chromosomes? 5
7. How many **DIFFERENT** alleles are represented? 7
8. Are T and d alleles for the same trait? No
9. The alleles of a person that is **homozygous dominant** for **eye color**: B B
10. The alleles of a person that is **homozygous recessive** for **Tay-Sachs disease**: t t
11. List the alleles for a person that is **heterozygous** for **dimples**: D d

Table 7-1. Key to Traits

Alleles	Trait
B, b	eye color
T, t	Tay-Sachs disease
D, d	dimples
A, a	presence of pigment melanin
E, e	attached or free ear lobes

ACTIVITY 7-2: RECESSIVE GENETIC DISORDERS

(application)

1. **Tay-Sachs disease** is a metabolic disorder that results in deterioration of the brain and nervous system, causing an **early death** in children (usually by age 5). The disease is caused by a **recessive allele (t)**. Stuart and Lauren's first child had Tay-Sachs disease. Lauren is pregnant again. State the genotypes of both parents and their first child. What are the chances their second child will have the disease? **Show all your work.**

   Genotypes:

   Stuart Tt   Lauren Tt   first child tt

   Probability of a second Tay-Sachs child: 1/4

	T	t
T	TT	Tt
t	Tt	tt

2. Flat feet are inherited through a **recessive allele (f)**. Two people who have normal arches produced a child with flat feet. What are the genotypes of all the family members? What is the probability their next child will have flat feet?

   Genotypes:

   father Ff   mother Ff   child ff

   Probability of another child with flat feet: 1/4

	F	f
F	FF	Ff
f	fF	ff

ACTIVITY 7-3: TRAITS INHERITED BY DOMINANT ALLELES

(application)

1. Kirk Douglas has a prominent dimple in his chin and so does his son Michael. Dimples are inherited by a **dominant allele (D)**. If Michael's mother does **not** have dimples, and Michael's wife does **not** have dimples, what is the probability that Michael's daughter will have dimples? **Show all your work.** Include the **genotypes** of all the people in the problem.

   Genotypes:

   Kirk ______ Michael ______

   Kirk's wife ______ Michael's wife ______

   Probability of Michael's daughter having dimples ________

2. Marfan syndrome is a dominant trait. **Dominant** traits are expressed even if **only one dominant allele is present (M)**. **Normal** individuals have **two recessive alleles (m)**.

   Chris Patton, a University of Maryland basketball player in the late 1970s, died while playing basketball. An autopsy showed that Patton's aorta had burst because of the lifelong weakening effects of Marfan syndrome. Assume that Chris's mother had Marfan syndrome and his father was normal. If Chris had married a normal woman, and had a child before his death, what is the probability that this child would inherit Marfan disease? **Show your work!**

   Genotypes:

   Chris's father ______ Chris's wife ______

   Chris's mother ______ Chris ______

   Probability of a Marfan syndrome child ______

ACTIVITY 8-1: POLYGENIC TRAITS

(application)

Polygenic traits are inherited through the interaction of several genes. Human traits thought to be inherited in this manner include skin and eye color, height, obesity, and intelligence. For purposes of illustration, the inheritance of skin color is assumed to be caused by the interaction of only two genes (A and B).

1. Using the information in **Table 8-1**, **draw a graph** in **Figure 8-1** showing the percent of people in the population who have inherited each of the five skin color shadings.

Table 8-1. Inheritance of Skin Color

Genotype	Phenotype	Number Showing Trait	Percent of Population
AABB	black	136	
AABb or AaBB	dark	267	
AaBb or AAbb or aaBB	medium	692	
Aabb or aaBb	light	273	
aabb	white	132	
TOTALS			

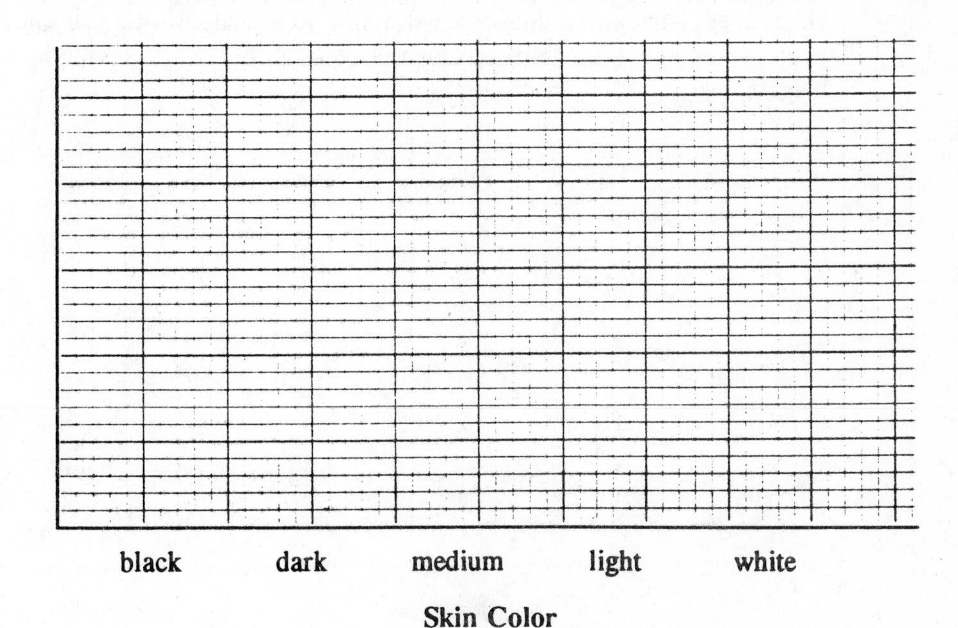

Figure 8-1. Frequency of Various Skin Colors

2. Give a genetic explanation for the fact that "medium" skin color appears more frequently in the population than any of the other shades.

3. If the **X** axis on your graph was changed to Height of Individuals, which **genotype** would produce the shortest group of individuals? **Explain your answer**.

Challenge Question!!

4. Give an explanation for the fact that height in humans shows a normal distribution of phenotypes, but pea plants produce only the tall or dwarf phenotype (no intermediate heights).

ACTIVITY 8-2: THE GENOTYPES AND PHENOTYPES OF BLOOD GROUPS

(comprehension)

The table below shows the blood types of mothers and children. Complete the table by filling in **all the possible genotypes** for the mother and child. For each given mother and child, list any genotypes **that can be eliminated as possibilities** for the father.

Mother Blood Type	Mother Possible Genotypes	Child Blood Type	Child Possible Genotypes	Impossible Genotypes for Father
A	$I^A I^A$ or $I^A i$	O	ii only	type AB
B	$I^B I^B$ or $I^B i$	AB	$I^A I^B$	B or O
O	$i i$	A	$I^A i$	B
AB	$I^A I^B$	A	$i i$	
O	$i i$	O	$i i$	A, B, AB
B	I^B	B	I^B	A
AB		AB		
A		AB		
B		A		
AB		B		

ACTIVITY 8-4: BLOOD TRANSFUSION CONFUSION

(application)

Kim's father, Alfred, is rushed to the hospital after a serious car accident. The doctor tells Kim and her mother that her father needs a blood transfusion. Kim offers to make the donation and they take her to the laboratory to check her blood type. Kim's blood type is AB. Her father's blood type is O.

Kim is a Biology student and she begins to wonder if she is adopted. What would you tell her? **Explain your answer.**

ACTIVITY 8-5: GENES ON THE X CHROMOSOME

(recall and comprehension)

i	I	Ichthyosis (dry scaly skin) - recessive
H	h	hypophosphatemia (Vitamin D resistant rickets) - dominant
D	D	Duchenne muscular dystrophy (progressive muscle weakness) - recessive
c	C	cleft palate (opening in roof of mouth)
G	g	agammaglobulinemia (lack of antibodies, no immune system defenses) - recessive
o	o	ocular albinism (no eye pigment)
L	l	Lesch-Nyhan syndrome (mental retardation, self mutilation) - recessive
f	F	fragile X (leads to mental retardation) - recessive
A	a	hemophilia A - recessive
e	e	red/green color blindness - recessive

Figure 8-2. Sections of Two X Chromosomes from a Human Karyotype

Answer the following questions by referring to the two chromosomes in **Figure 8-2**. The chromosomes have been mapped to determine which alleles they carry.

1. For how many traits is this person **homozygous dominant?** ______

   For how many traits is this person **heterozygous?** ______

2. Does this pair of chromosomes belong to a **male or a female?** ____________

3. What is this person's **phenotype** for the red/green color blindness trait? __________

   What are the most likely genotypes of both parents?

   mother: ________ father: ________

   What is the probability that this person will have a **son** that is also color blind? ______

4. Imagine that, during gamete formation, nondisjunction occurred between these two chromosomes. Which of the following chromosome combinations would be possible in the offspring? (Circle **all** answers that apply.) **Explain your answer.**

   a. XYY
   b. XO
   c. XXY
   d. XXX
   e. XY
   f. XX

Challenge Question!!

5. Which sex linked traits could be an **exception** to the normal rule, which states that sex linked traits are more commonly expressed in males than females? **Explain your answer**.

ACTIVITY 8-6: SEX LINKED TRAITS

(application)

1. Both the mother and father of a hemophiliac son (a sex linked, recessive trait) have normal blood clotting. How did the son inherit the hemophilia gene?

   What are the genotypes of the mother, the father, and the son?

2. A woman is color blind, a sex linked recessive trait. If she marries a man with normal vision, what are the chances her sons will be color blind? What are the chances her daughters will be color blind? What is the chance of having a carrier daughter?

3. Both the husband and wife have normal vision. Their daughter is color blind. What can you conclude about the girl's father?

4. If you remove a cell from the inside of your cheek or your elbow, how many sex chromosomes would you find inside each of these cells? **Explain your answer.**

ACTIVITY 8-7: MORE PRACTICE WITH SEX LINKED TRAITS

(application)

1. If a young girl has fragile X syndrome, a sex linked, recessive trait (**f**), what is her genotype? __________

   What are the most probable genotypes of her father and mother?

   father: __________ mother: __________

   If her brother also developed this condition, which parent (father, mother, or both) contributed a disease allele? **Explain your answer.**

2. Two parents gave birth to a child with ichthyosis, a sex linked, recessive trait. If one of the **mother's** X chromosomes carried the **dominant allele (D)**, what allele must be present on her other X chromosome? ______

   If an X chromosome taken from the child carries the **recessive allele (d)**, what allele must be present on the other sex chromosome?

   female child: ______ male child: _______

ACTIVITY 9-1: DNA STRUCTURE

(recall)

1. Fill in the shaded boxes to complete this DNA molecule. Use the **symbol P** for **phosphate**, S for five carbon **sugar**, and A, T, C, and G for the appropriate nitrogen **bases**.

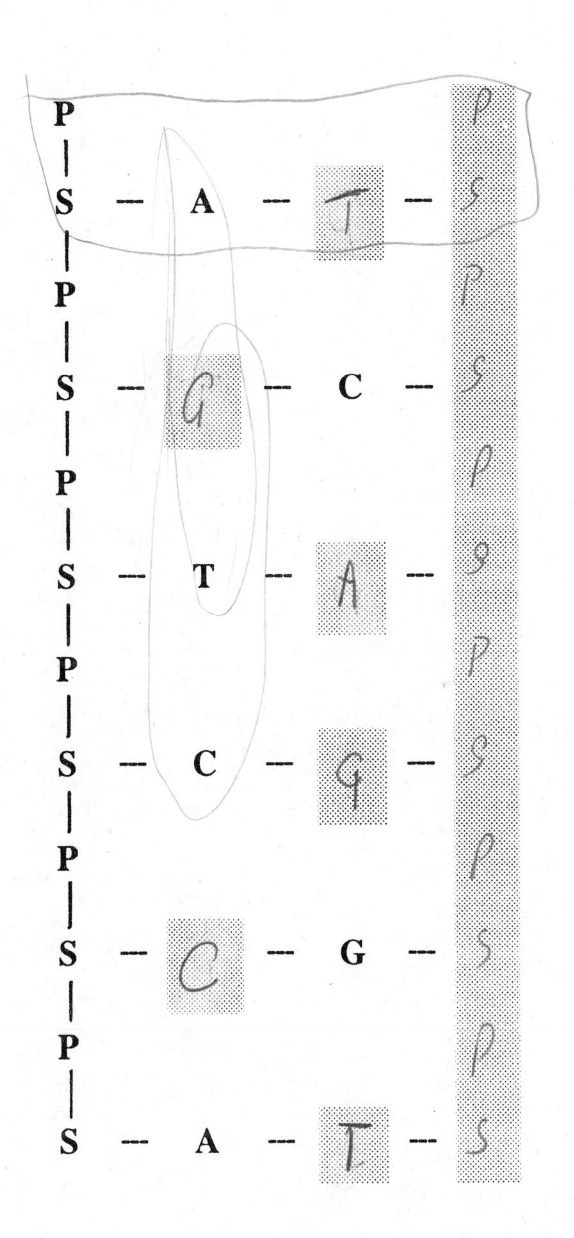

2. List the names of the four nitrogen bases included in a DNA molecule.

   adenine, cytosine, Guanine, tymine

3. Draw a box around **one** complete nucleotide.

4. What is a DNA triplet?

   complementary three base sequence on the dna

5. How many triplets are on the **left** side of this DNA section?

   2

6. Which cellular organelle would contain this DNA molecule?

   nucleus

7. What is the **base pairing rule**?

   it holds two strands of DNA together.

   ATCG

ACTIVITY 9-2: USING RADIOACTIVE LABELING TO TRACE CELL ACTIVITIES

(recall)

Human pancreatic cells are immersed in a radioactively labeled nutrient broth for several hours and then removed. After removal, the cells were monitored for traces of radioactivity in various organelles.

Thirty minutes after removal from the labeled nutrient broth, the cells showed radioactivity in the rough ER. After 90 minutes, the radioactivity was present in the smooth ER, at two hours in the Golgi bodies, and a half hour later in vesicles near one end of the cell. After four hours had expired, no radioactivity could be detected within any of the cells.

1. What cellular process has this experiment traced? **Explain your answer.**

Challenge Question!!

2. If this process was occurring inside your body, what molecule(s) might have been released into the bloodstream by the activity of these pancreatic cells?

ACTIVITY 9-3: CONNECTING mRNA WITH DNA

(comprehension)

Answer the following questions in reference to **Table 9-1**.

Table 9-1. mRNA Codons and Their Corresponding Amino Acids

UUU	phenylalanine	UCU	serine	UAU	tyrosine	UGU	cysteine
UUC	phenylalanine	UCC	serine	UAC	tyrosine	UGC	cysteine
UUA	leucine	UCA	serine	UAA	terminator	UGA	terminator
UUG	leucine	UCG	serine	UAG	terminator	UGG	tryptophan
CUU	leucine	CCU	proline	CAU	histidine	CGU	arginine
CUC	leucine	CCC	proline	CAC	histidine	CGC	arginine
CUA	leucine	CCA	proline	CAA	glutamine	CGA	arginine
CUG	leucine	CCG	proline	CAG	glutamine	CGG	arginine
AUU	isoleucine	ACU	threonine	AAU	asparagine	AGU	serine
AUC	isoleucine	ACC	threonine	AAC	asparagine	AGC	serine
AUA	isoleucine	ACA	threonine	AAA	lysine	AGA	arginine
AUG	start (methionine)	ACG	threonine	AAG	lysine	AGG	arginine
GUU	valine	GCU	alanine	GAU	aspartic acid	GGU	glycine
GUC	valine	GCC	alanine	GAC	aspartic acid	GGC	glycine
GUA	valine	GCA	alanine	GAA	glutamic acid	GGA	glycine
GUG	valine	GCG	alanine	GAG	glutamic acid	GGG	glycine

1. There are __4__ **DNA triplets** that all specify the same amino acid: **valine.**

   Each triplet differs from the others that specify valine by __1__ nucleotide(s).

2. What **DNA triplet** was used as the **template** for the last mRNA codon for valine (**GUG**)?

   CAC

3. If there was a **point mutation** in a DNA triplet that changed the code from **T T G** to **T T A**, would that cause a problem with the resulting polypeptide chain? **Explain your answer**.

No, same amino acid

4. If a different point mutation changed the DNA code from **A C G** to **A C T**, would that cause a problem with the resulting polypeptide chain? **Explain your answer**.

Yes, because it's gonna stop the protein from synthesis.

ACTIVITY 9-6: USING DNA FINGERPRINTS

(application, analysis, synthesis, and evaluation)

As a result of years of war and conflict in a South American country, many children have been separated from their families. Many of these children are living in orphanages run by the United Nations and the International Red Cross. After several years, an attempt is being made to return these children to their parents. DNA profiles of two sets of parents who are in search of missing children were compared with the profiles of two girls of the same age who are believed to have been taken from the same village.

1. Are either of the children (1 or 2) good genetic matches for Julio and Maria?

   Explain your answer, citing allele matches from specific loci of their DNA profiles.

2. Are either of the children (1 or 2) good genetic matches for Ernesto and Eva?

   Explain your answer, citing allele matches from specific loci of their DNA profiles.

		L	Julio	Maria	Ernesto	Eva	Child 1	Child 2	L	
Locus A	1	--	-------						--	1
	2	--		-------	-------		-------	-------	--	2
	3	--			-------				--	3
	4	--	-------			-------	-------	-------	--	4
	5	--				-------			--	5
	6	--							--	6
Locus B	1	--							--	1
	2	--		-------	-------		-------		--	2
	3	--	-------	-------			-------		--	3
	4	--	-------		-------	-------		-------	--	4
	5	--				-------			--	5
	6	--							--	6
Locus C	1	--		-------		-------	-------		--	1
	2	--	-------			-------	-------	-------	--	2
	3	--			-------				--	3
	4	--		-------					--	4
	5	--							--	5
	6	--			-------			-------	--	6
	7	--	-------						--	7
Locus D	1	--							--	1
	2	--		-------		-------	-------		--	2
	3	--	-------			-------	-------	-------	--	3
	4	--			-------			-------	--	4
	5	--		-------					--	5
	6	--							--	6
	7	--							--	7
Locus E	1	--	-------						--	1
	2	--			-------				--	2
	3	--			-------			-------	--	3
	4	--		-------		-------			--	4
	5	--						-------	--	5
	6	--	-------	-------			-------		--	6
Locus F	1	--	-------			-------	-------		--	1
	2	--	-------		-------			-------	--	2
	3	--							--	3
	4	--		-------		-------	-------		--	4
	5	--			-------			-------	--	5
	6	--		-------					--	6
Locus G	1	--							--	1
	2	--	-------		-------		-------		--	2
	3	--				-------			--	3
	4	--		-------	-------	-------	-------	-------	--	4
	5	--							--	5
Locus H	1	--		-------		-------	-------		--	1
	2	--	-------		-------		-------		--	2
	3	--				-------			--	3
	4	--		-------				-------	--	4
	5	--			-------				--	5
	6	--	-------						--	6

Figure 9-2. DNA Profiles

ACTIVITY 9-7: BIGFOOT OR BIG HOAX?

(application, analysis, and synthesis)

You are a scientist on an expedition to find Bigfoot in the American northwest. Your team finds a huge set of footprints in the woods and you make plaster casts for later study. While following the tracks, you also collect some hair from a tree branch, blood from a barbed wire fence, and several bones from a cave.

What steps would you follow to determine whether any of the collected samples might possibly belong to Bigfoot? **Be specific. Be complete.**

ACTIVITY 10-1: EVOLUTION OF PESTICIDE RESISTANCE

(comprehension)

Since pesticides were introduced to agriculture in the early 1940s, many pest species have evolved resistance to specific chemicals. This is a result of a type of directional selection that occurs in response to new environmental conditions. The frequency of alleles that confer pesticide resistance increase steadily in the population.

The graph in **Figure 10-1** illustrates the trend in pesticide resistance among arthropod crop pests (insects and mites).

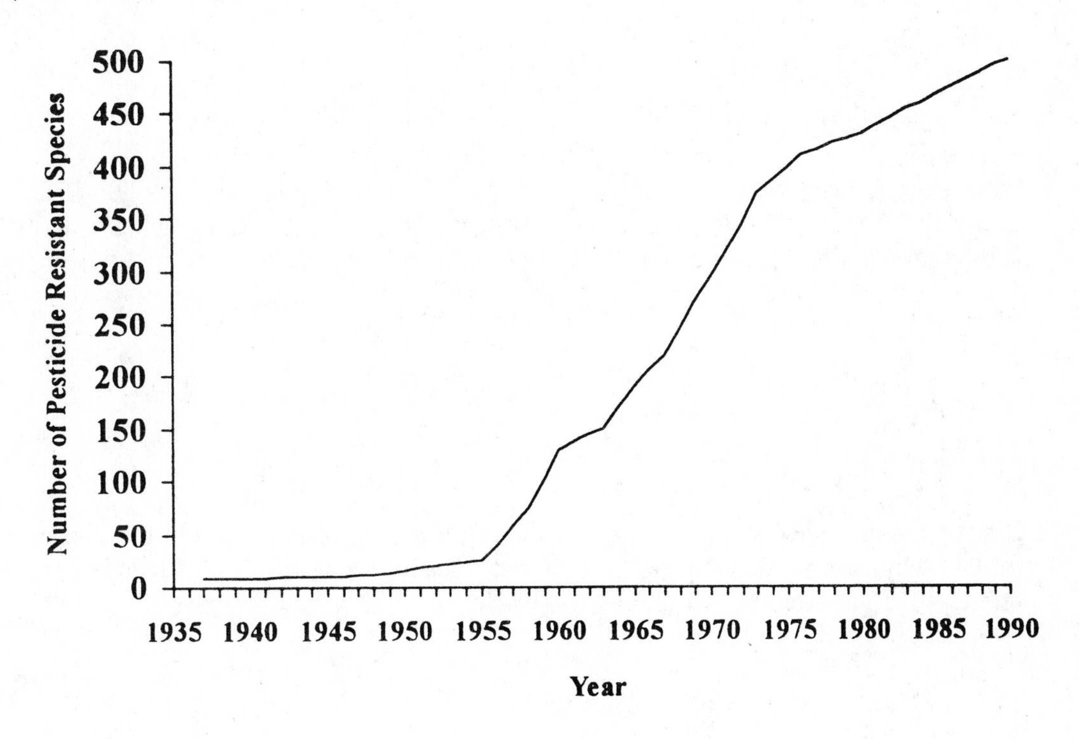

Figure 10-1.

1. Select the best answer. According to the data **presented in the graph**:

   a. each pest species will gradually become more and more resistant to a particular pesticide
   b. an insect can become more and more resistant to pesticides during its lifetime
   c. insecticide use was less common in 1990 compared to 1960
   d. more and more insect species are developing pesticide resistance
   e. use of pesticides sharply increased after World War II

2. The graph in **Figure 10-1** is missing its title. Based on your understanding of the information presented, write in an appropriate title for this graph.

3. Calculate the **average** number of resistant species between **1955 - 1965**. Compare this number with the average number of resistant species between **1980-1990**. Summarize your findings in one or two sentences.

4. Imagine that a new pesticide has been developed and is introduced in 1999. If the evolution of resistance followed the same pattern as that shown in the graph above, in approximately what year would there be 500 resistant species?

Challenge Question!!

5. Based on your understanding of the evolution of pesticide resistance in insect species, explain why treating tuberculosis (a bacterial infection) with antibiotics today is less effective than in previous decades.

ACTIVITY 10-2: INFERRING EVOLUTIONARY RELATIONSHIPS FROM AMINO ACID SEQUENCES

(application)

Although bacteria, worms, and humans appear to have little physical similarity, they are quite similar in terms of the proteins that control metabolic activities. Proteins that are found in different organisms contain many of the same amino acid sequences. Since proteins are coded in DNA molecules, organisms with similar proteins must have similar DNA. The greater the similarity of nucleotide sequences within these proteins, the more closely two species are related. Scientists studying evolution have found that charting biochemical similarities produces family trees that are similar to those constructed using other lines of evidence.

One molecule commonly studied to determine evolutionary relationships is cytochrome C, a protein that plays an important role in cell respiration. Cytochrome C has a primary structure of 104 amino acids.

Answer the following questions in reference to **Table 10-1**.

Table 10-1. Number of Amino Acid Differences in Cytochrome C Between Humans and Other Species

Species	Differences	Species	Differences
fruit fly	29	rhesus monkey	1
moth	31	pig	10
pigeon	12	dog	11
chimpanzee	0	kangaroo	10
rattlesnake	14	wheat	43
horse	12	infectious yeast (*Candida*)	51
red bread mold (*Neurospora*)	48	brewer's yeast (*Saccharomyces*)	45

1. On the basis of differences in the amino acid sequence of cytochrome C, which organisms appear to be **most closely** related to **humans**?

2. Which organisms appear to be **most distantly** related to humans?

3. One celled organisms, such as bacteria and yeast, may have up to 50% difference in the amino acid sequence of their cytochrome C molecules compared to humans. Since humans and microorganisms are related to each other through common descent, how could their cytochrome C structures have become so different?

4. Other proteins, such as hemoglobin, are also used to establish degrees of evolutionary relationship. If you were to compare the hemoglobin of chimpanzees and humans, approximately how many amino acid differences would you expect to find? **Explain your answer**.

   a. zero to five differences
   b. five to ten differences
   c. more than ten differences

ACTIVITY 10-3: WHAT HAPPENED TO THE ALLIGATORS?

(comprehension)

A Florida theme park began a new promotion to attract visitors from New England and the Midwest. Free alligators were given away to residents of these two areas if they booked their trip between March and October and stayed at one of the resort hotels. Thousands of families took advantage of this offer, carried their baby alligators back home, and raised them over the winter. By the following spring, most of the "adopted" alligators had grown too large to keep at home. Many parents decided that the nearest river would provide a perfect habitat and released their alligators to the wild. Several years later, the expected proliferation of alligators had not occurred. The rivers were as free of alligators as they had been before the pets were released.

Local newspapers presented the following three explanations about the fate of the "missing" alligators. Determine whether **any** of these three explanations are valid and can be supported by principles of evolutionary change and extinction. **Explain why you consider each of the statements to be valid or invalid.**

1. This is an example of **directional selection**. Directional selection, which occurs when members of a species migrate to a new habitat with different environmental conditions, acted against those alligators most sensitive to cold weather. Changes in the gene pool led to the formation of a new "improved" species of alligator that was adapted to local conditions. They migrated up and down the rivers establishing territories in new geographic locations around New England and the Midwest.

2. **Punctuated equilibrium** is responsible for the observed lack of alligators. Speciation occurred during the first winter, so that by spring, the isolated alligators had evolved into terrestrial forms that foraged in nearby wooded areas.

3. The alligators were not able to **adapt** to their new environment. Extreme weather changes occurred too rapidly for evolutionary change to occur. The released individuals were killed by the winter cold.

ACTIVITY 10-4: SAVING ENDANGERED SPECIES

(comprehension)

The current population of gorillas in the wild is quite low. Less than 100 captive gorillas, descended from only a few pairs of wild caught parents, are living in zoos around the world. Assume that funding and space were available for a captive-breeding program for the purpose of releasing several hundred offspring from the zoo population to replenish the endangered wild groupings.

1. From an **evolutionary perspective**, what would be the effect of this plan on the **genetic variability** of the released individuals?

2. What effect might the **level of genetic variability** of released individuals have on the **long-term survival** of the wild gorilla population? (**Note:** Consider your answer to the previous question.)

ACTIVITY 11-1: OVERVIEW OF FEEDBACK CONTROL

(comprehension)

Circle the best answer to each of the following multiple choice questions.

1. Which of the following is an example of positive feedback?

   a. an increase in blood sugar concentration increases the amount of the hormone that causes sugar to be stored as glycogen
   b. a decrease in blood sugar concentration increases the amount of the hormone that triggers the conversion of glycogen to glucose
   c. an infant's sucking at the mother's breast increases the amount of the hormone that induces the release of milk from the mammary glands
   d. an increase in calcium concentration increases the amount of the hormone that causes calcium storage in bone
   e. a decrease in calcium concentration increases the amount of the hormone that causes the release of calcium from bone

2. Which of the following statements best illustrates the concept of homeostasis?

   a. the pH of the blood is maintained by an internal buffer system
   b. the evolution of human beings has been relatively static
   c. water is a polar molecule and thus forms hydrogen bonds
   d. green plants display more diversity than fungi
   e. all of the above statements illustrate homeostasis

3. Circle examples of negative feedback control/homeostasis:

   a. as body temperature rises, sweating occurs
   b. when a person drinks large quantities of water, the kidneys produce more urine
   c. when a person eats large amounts of salt, the kidneys excrete more salt
   d. when body temperature rises, more blood flows to the skin
   e. when a person is exercising heavily, the heart beats more slowly

Challenge Question!!

4. You ate a banana for your mid-morning snack. The potassium in the banana increases the potassium concentration in your blood. According to the principles of **HOMEOSTASIS**, how should your body respond? **Explain your answer**.

ACTIVITY 11-2: NEGATIVE FEEDBACK CONTROL OF BLOOD GLUCOSE LEVELS

(comprehension)

Refer to **Figure 11-1** and answer the following questions.

1. What organ produces insulin and glucagon? ________________

2. Which hormone is responsible for raising blood glucose levels? ________________

3. Which hormone is responsible for reducing blood glucose levels? ________________

4. Which hormone is released when the body needs more ATP? ________________

5. Which hormone is released after a large candy bar passes through the small intestine?

   ________________ **Explain your answer.**

6. What is the liver's response to the release of insulin?

7. What is the liver's response to the release of glucagon?

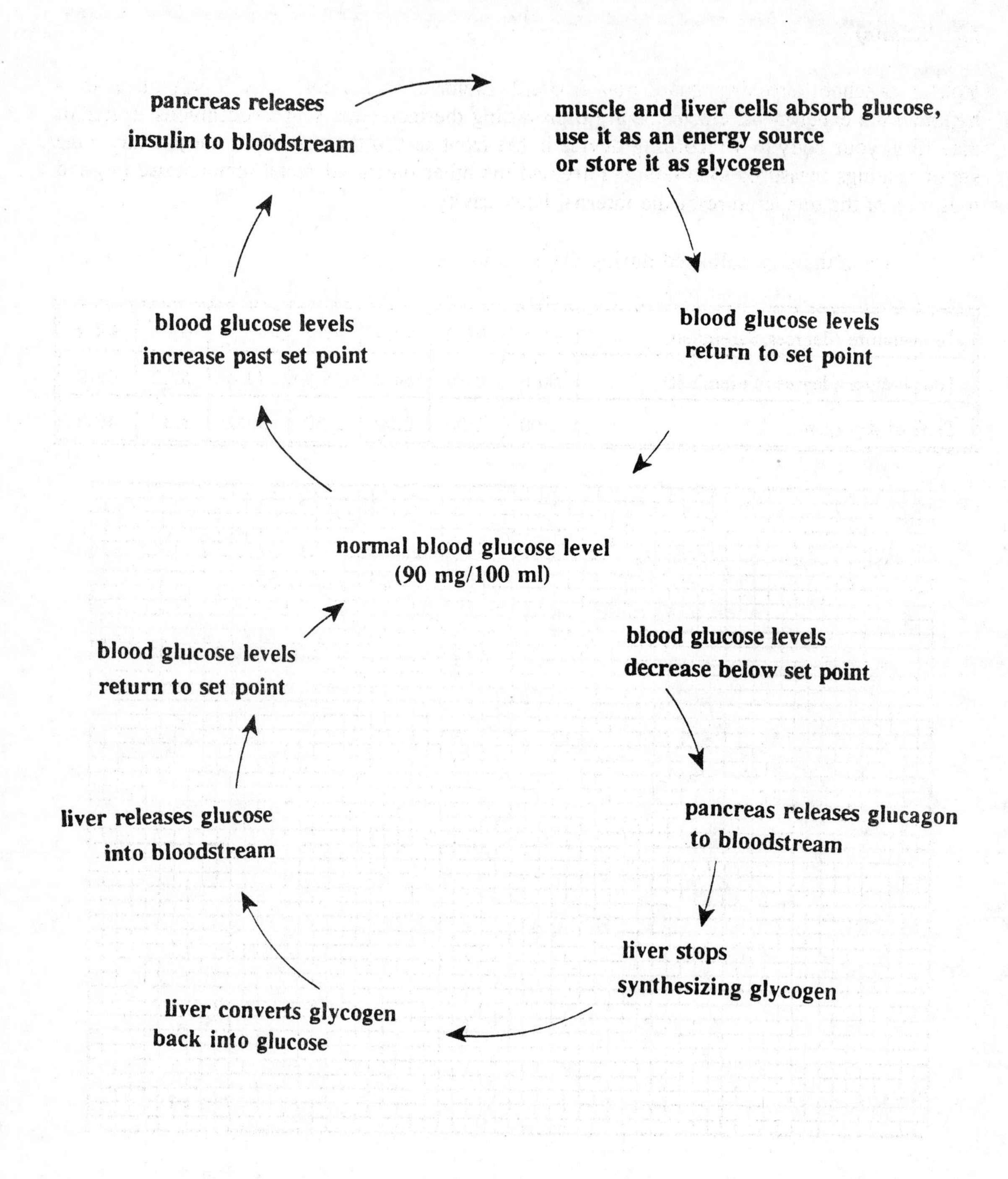

Figure 11-1. Negative Feedback Control of Blood Glucose Levels

ACTIVITY 11-3: COLD HANDS, WARM HEART

(application)

You're a school crossing guard for Howard Elementary School. As a volunteer in a hypothermia experiment, electronically transmitting thermometers sent a continuous stream of data from your body to a recording device in the front seat of the scientist's parked car. One set of readings measured skin temperature and the other measured rectal temperature (a good indicator of the temperature of the internal body cavity).

1. Graph the data collected during the experiment.

Temperature (degrees Fareinheit)	98.8	98.2	97.9	97.4	97.7	98.3	98.8
Temperature (degrees Fareinheit)	80.0	81.0	84.0	75.5	70.4	69.2	68.0
Time of day (p.m.)	1:00	1:30	2:00	2:30	3:00	3:30	4:00

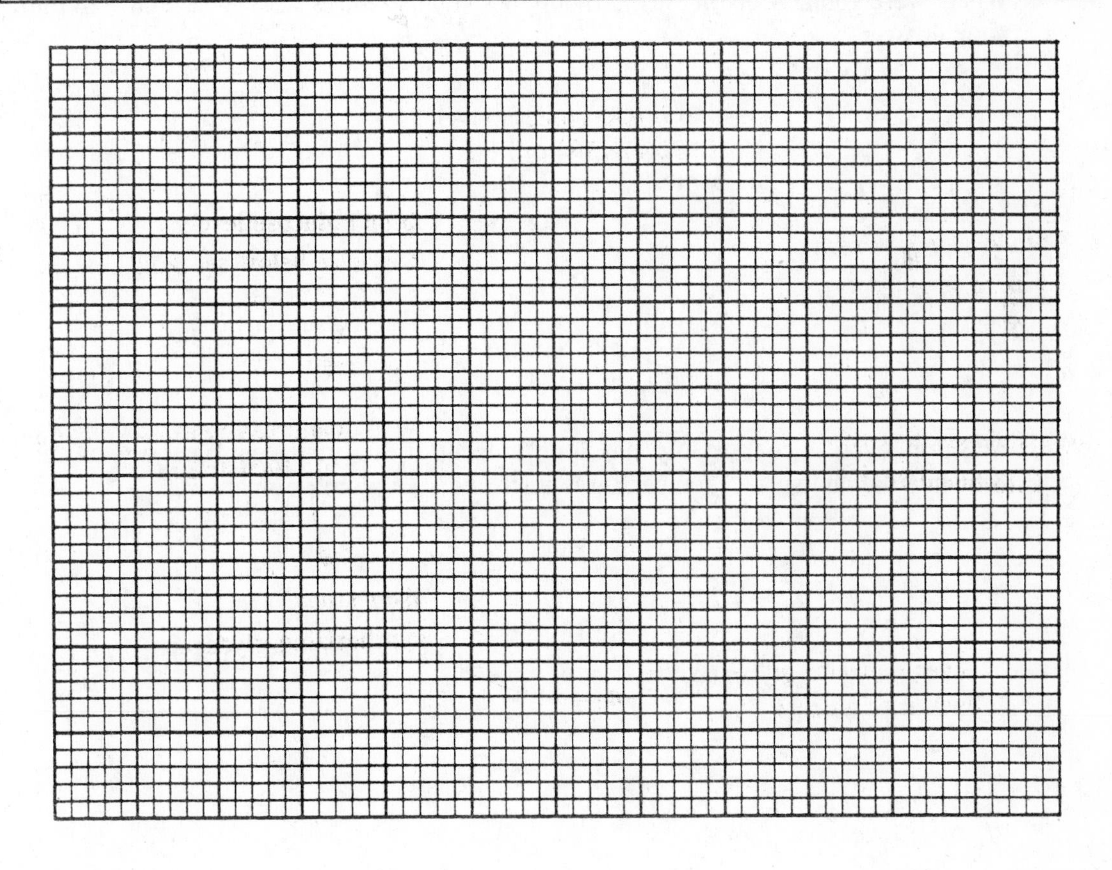

Figure 11-2.

2. Use your **critical reasoning skills** to determine which lines on your graph should be labeled "core temperature" and which should be labeled "skin temperature." **Label** the lines and give the graph an appropriate **title**.

3. What time of day did your crossing guard shift begin? **Explain your answer**.

4. Where were you at 4:00 on the afternoon these temperatures were measured? **Explain your answer**.

5. At 8 p.m., you're comfortably curled up in front of the fireplace. What do you think your internal and surface temperatures might be? **Explain your answer**.

ACTIVITY 11-4: INTERPRETING GLUCOSE TOLERANCE TESTS

(application)

One method of diagnosing diabetes mellitus is a glucose tolerance test. The patient drinks a concentrated glucose solution and her blood-glucose level is monitored for several hours. Answer the following questions in reference to **Figure 11-3**, which shows the results of a glucose-tolerance test.

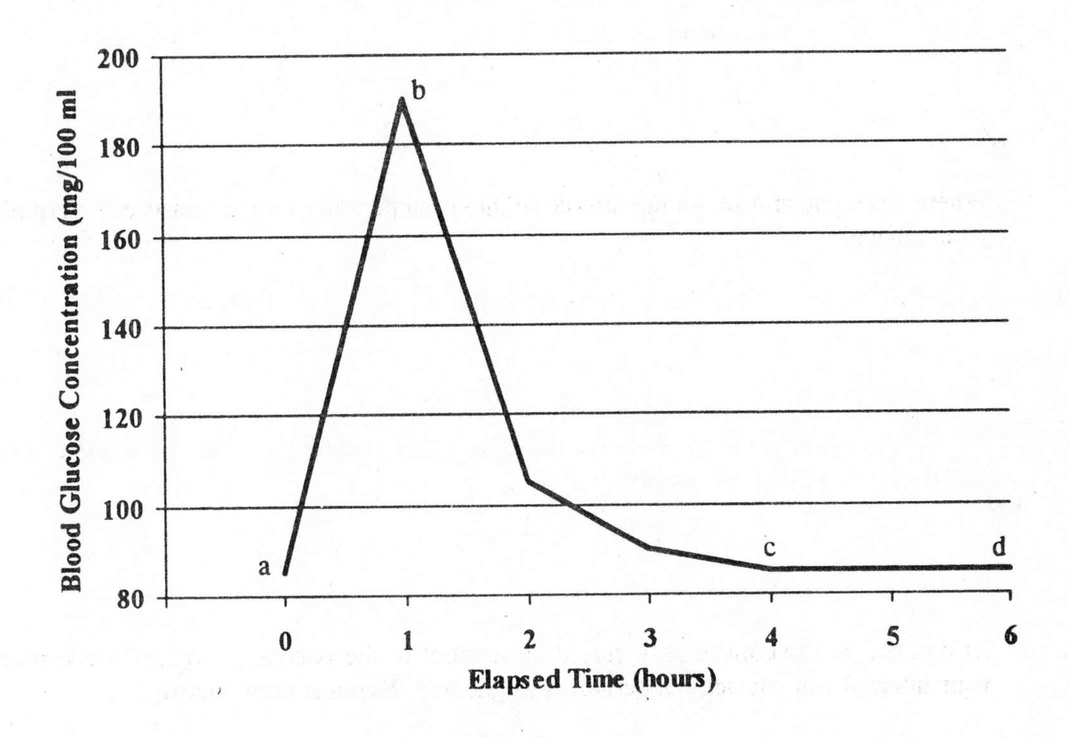

Figure 11-3. Results of Glucose Tolerance Test for Patient A

1. What is the body's **set point** for blood glucose? How did you arrive at this answer? **Explain completely.**

2. What is happening in this patient's body between **letters a and b** on the graph?

3. What is happening between **letters b and c** on the graph?

4. Briefly summarize the steps occurring in the body that can account for the observed change between **b and c** on the graph.

5. If you were this patient's doctor, would you feel that medication for diabetes was justified? **Explain your answer**.

Challenge Question!!

6. If your answer to question #5 was **yes**, alter the line on the graph to show a person **without** diabetes.

   If your answer to question #5 was **no**, alter the line to show the results of a glucose tolerance test for a **diabetic**.

HW #: 12-4, 16-4

ACTIVITY 12-1: DIGESTIVE SYSTEM BASICS

(recall)

Use **Figure 12-1** to answer the following questions. Answers can be used **more than once**.

1. 5 Which number represents an organ in which **most** chemical digestion by enzymes will occur?

2. 7 This organ is important for **glycogen** storage.

3. 2 This organ is used primarily to **transport** food.

4. 3 This organ **stores** food. Chemical and mechanical digestion also take place here.

5. 9 Reabsorption of water is the main function of this organ.

6. 5 The first part of this organ contains ducts which allow secretions from the pancreas and liver to enter the digestive system.

7. 8 This organ **stores** a liquid that emulsifies fat.

8. 1 These glands produce an enzyme responsible for the **initial** breakdown of starch.

9. 4 Sodium bicarbonate released from this organ neutralizes acid in the small intestine.

10. 7 This organ produces bile.

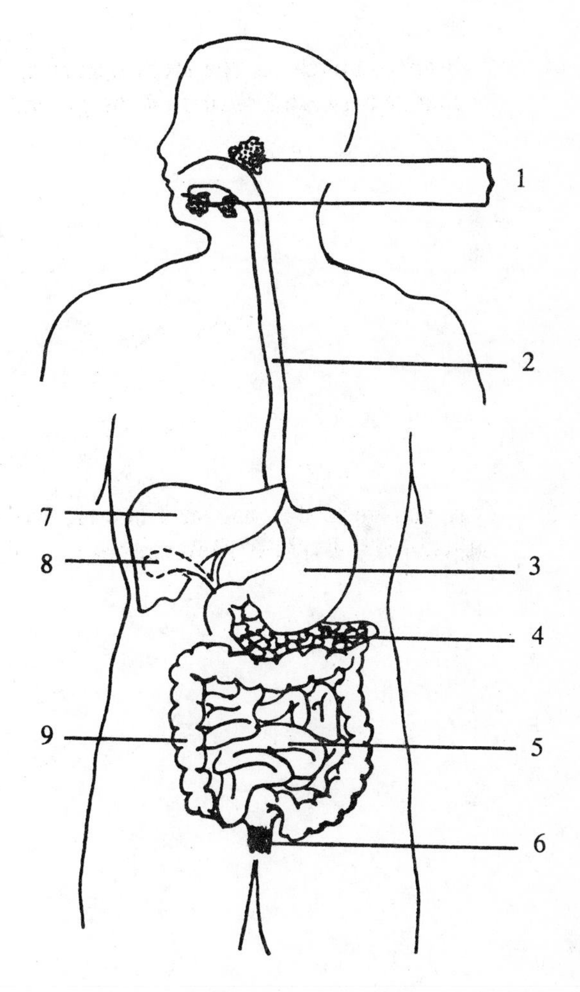

Figure 12-1. Human Digestive System
From Human Biology: Condensed, by Bres and Weisshaar, p. 2-49. Copyright (©) 1997 (Education Resources). Reprinted by permission.

ACTIVITY 12-2: DIGESTIVE ENZYMES

(recall)

Fill in the blanks in the following chart of digestive enzymes.

Enzymes	Secreted By	Functions In	Hydrolyzes
salivary amylase	mouth		starch to maltose (a disaccharide)
pepsin		stomach	
	pancreas	small intestine	fats to glycerol and fatty acids
trypsin	pancreas		
pancreatic amylase	pancreas	small intestine	
peptidase			dipeptides to two amino acids
sucrase (also called invertase)	small intestine	small intestine	
	small intestine	small intestine	maltose to two glucose molecules
	small intestine	small intestine	lactose to glucose and galactose

ACTIVITY 12-3: DIETARY FAT COMPARISON

(application)

Read the dietary fat graph in **Figure 12-2** and answer the following questions.

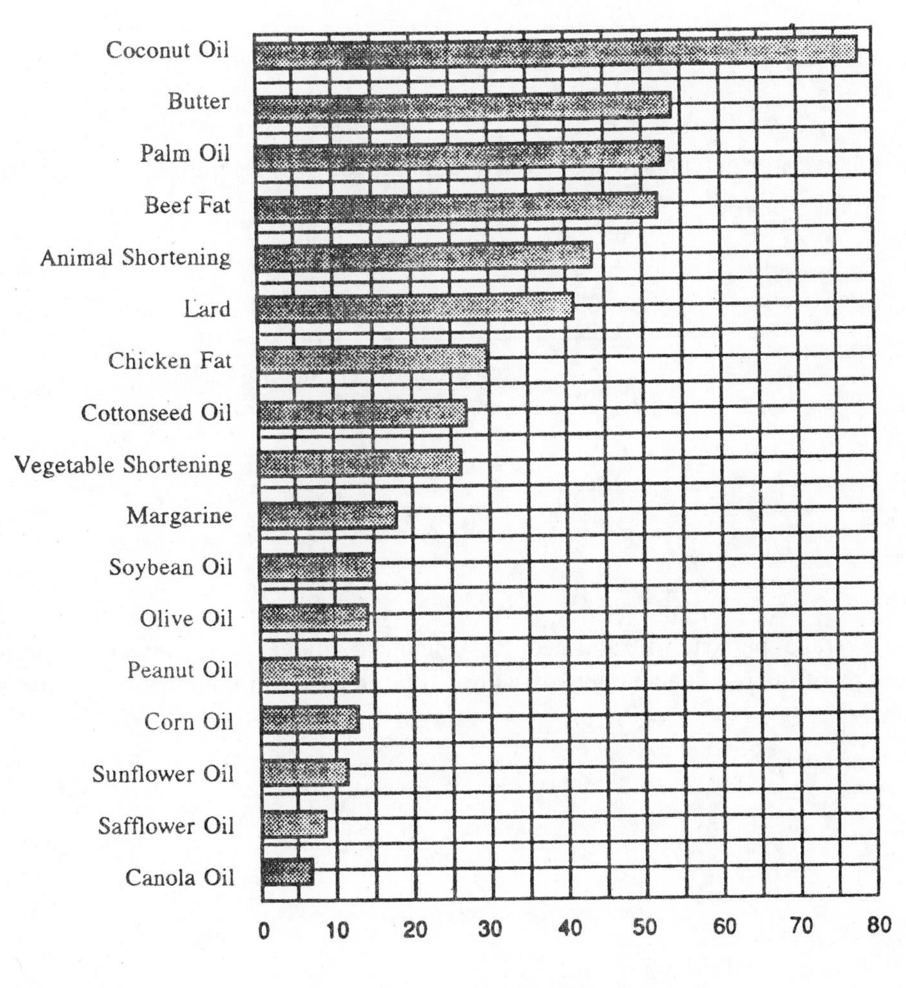

Figure 12-2. Comparisons of Dietary Fat

1. What percent of the total fat in butter is saturated? _____

2. A friend who is watching her diet told you that vegetable shortening has only **half** the saturated fat contained in butter. Is she correct? **Explain your answer.**

3. Calculate the **average** percentage of saturated fat in the following common cooking oils: corn oil, canola oil, safflower oil, and coconut oil.

$$\textbf{Average} = \frac{\text{total of the saturated fat in all four oils}}{\text{number of different oils}}$$

Average = ____________

4. Does the average percent of saturated fat calculated above give you an accurate representation of the relative amounts of saturated fat in each of the four oils? **Explain your answer.**

ACTIVITY 12-5: VITAMINS AND MINERALS

(comprehension)

Choose yes or no for each item below.

Key Points	Vitamins	Minerals
are organic compounds	YES NO	YES NO
are inorganic compounds	YES NO	YES NO
contain carbon	YES NO	YES NO
have covalent bonds	YES NO	YES NO
function as "helpers" for enzyme reactions	YES NO	YES NO
may be water or fat soluble	YES NO	YES NO
may become part of body tissues	YES NO	YES NO
if overdosed, can accumulate in body tissues to toxic levels	YES NO	YES NO
some molecules in this group have been identified as "essential" to the human diet	YES NO	YES NO

Hlw 6 = 13-3, 13-4, 14-1, 14-6

ACTIVITY 13-1: RESPIRATORY REVIEW

(recall)

Choose the most appropriate answer for the questions below. Answers can be used **more than once**.

A.	**LARYNX**	**F.**	**EPIGLOTTIS**	**K.**	**BRONCHIOLES**
B.	**GLOTTIS**	**G.**	**BRONCHII**	**L.**	**DIAPHRAGM**
C.	**TRACHEA**	**H.**	**HARD PALATE**	**M.**	**PLEURA**
D.	**ALVEOLI**	**I.**	**ESOPHAGUS**	**N.**	**VOCAL CORDS**
E.	**RIBS**	**J.**	**STOMACH**	**O.**	**NOSTRILS**

1. K smallest respiratory passageways
2. h partition that separates the nasal and oral cavities
3. C large tube that carries air to the lungs
4. C another name for "windpipe"
5. F prevents food from entering the larynx when swallowing
6. D area of the lungs where oxygen is absorbed and carbon dioxide is released
7. O first structure through which air passes when entering the body
8. L when this structure contracts, inhalation occurs
9. E provides protection for the organs of the thoracic cavity
10. M produces lubricating fluid that reduces friction when the lungs expand

ACTIVITY 13-2: BREATHING IN AND OUT

(comprehension)

1. In reference to **Figure 13-1**, label the bell jar model of the chest cavity with the name of each anatomical part being simulated.

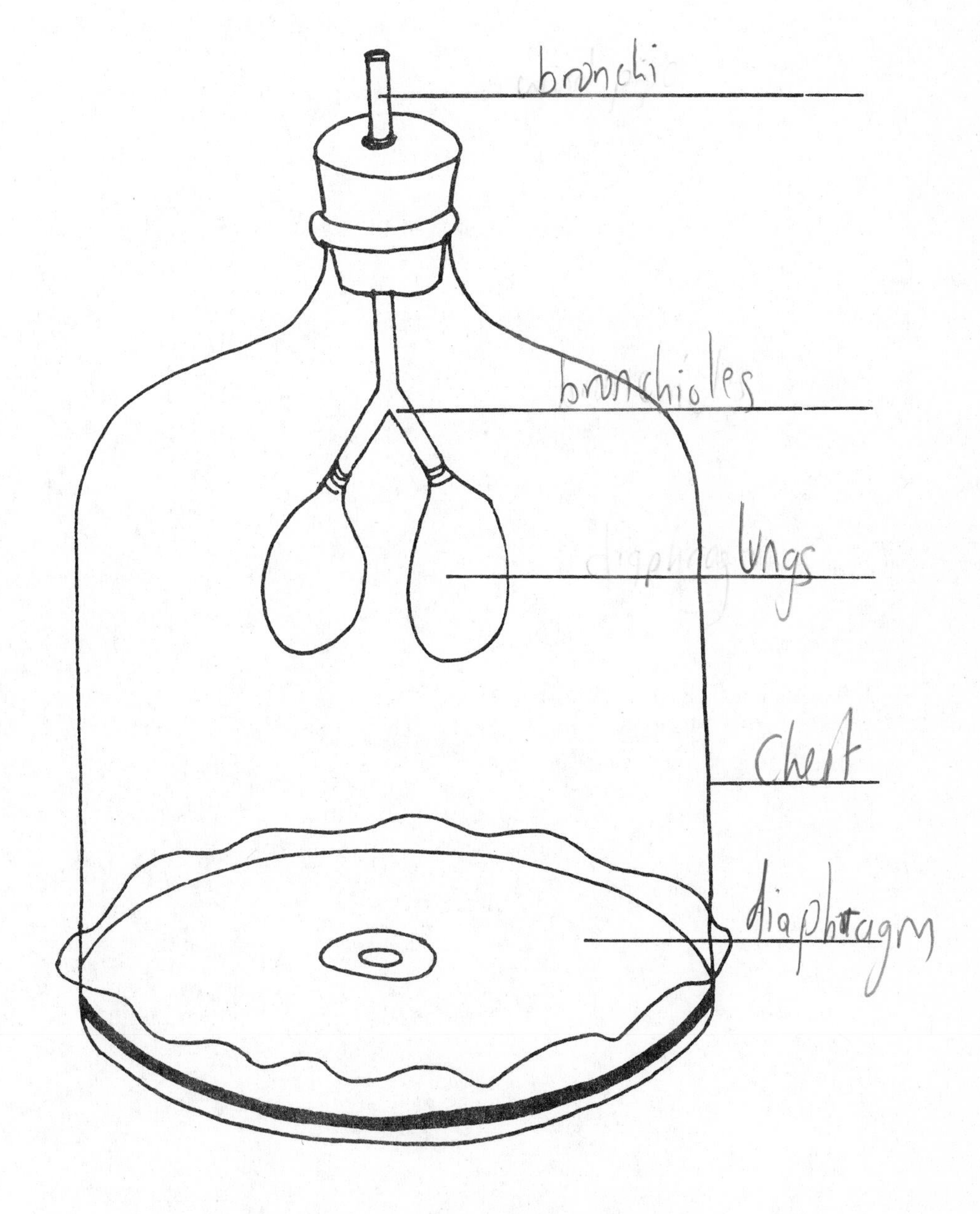

Figure 13-1. Bell Jar Model of the Chest Cavity

2. If you were to pull on the knob attached to the rubber sheet fastened across the bottom of the chest cavity model:

   a. Would you be simulating **inhalation or exhalation?** ________________

   b. What effect would pulling the knob have on **air pressure** in the chest cavity?

      decrease

   c. What would happen to the **size** of the two **balloons** in the model? **Why?**

      Contract

3. If you **release** the force pulling on the rubber sheet, so the sheet moves upward and inward, what will happen to the **size** of the two **balloons** in the model? **Explain** what **causes** the balloons to change in size.

ACTIVITY 13-5: AUTOPSY EVIDENCE LINKED TO THE RESPIRATORY SYSTEM

(application)

You are the police chief of a small town. You are investigating the recent death of an abandoned infant. The parents maintain that the baby was stillborn, but you suspect that the baby was born healthy. You are trying to collect enough evidence to determine whether a murder was committed.

The following evidence was cited in the pathologist's report:

a. When placed in a container of water, the lungs floated. (**Note:** Once lungs have been inflated, they will contain some residual air, even when collapsed.)

b. Significant blood was present in the vessels of the pulmonary circuit.

c. The ductus arteriosus and foramen ovale are closed.

Has a murder been committed? Explain how **each** piece of evidence above substantiates your decision.

ACTIVITY 14-2: BLOOD VESSEL DIAMETER AND BLOOD FLOW

(application)

Several students performed an experiment to determine the effect of blood vessel diameter on the rate of blood flow. Blood vessels were simulated with plastic tubes of varying diameters and the blood was simulated with tap water.

This experiment considered **only** the effect of vessel diameter, so the fluid was moved **only** by the force of gravity. Water was released into each tube for 30 seconds and collected in a graduated cylinder. The results of the experiment are summarized in **Table 14-1**.

Table 14-1. Results of Blood Flow Experiment

Tube Diameter (inches)	Volume Collected (ml/30 sec)
5/16	1780
4/16	1425
3/16	850
2/16	430

1. In one sentence, **summarize** the results of this experiment.

2. Considering that the above tube diameters are being used to **simulate** blood vessels, which diameter would be analagous to the aorta?

   Which diameter could represent a capillary?

3. From the results of the experiment, predict the effect of **vasoconstriction** on the **rate** of blood flow.

4. You have been out pulling weeds in your garden in the blazing summer sun. Your daughter hugs you and comments that your face feels hot. Explain how this heat is caused by changes in the **rate** of blood flow and **blood vessel diameter**.

ACTIVITY 14-3: REVIEW OF CIRCULATORY SYSTEM BASICS

(recall)

Fill in the blank with the most appropriate answer. Answers can be used **more than once**.

1. RBC — type of blood cell that plays a role in the respiratory system

2. fibrin — holds platelets in position over an injury

3. plasma — comprises the largest percentage of the blood volume

4. WBC — type of blood cell that plays a role in the immune system

5. Red — most numerous type of blood cell

6. hemoglobin — lack of this protein will decrease oxygen supply to body tissues

7. Veins — carries blood towards the heart

8. veins — blood vessel containing one-way valves

9. lipoids — may become swollen due to an infection

10. plasma — part of the blood which transports salts, glucose, and amino acids

ACTIVITY 14-4: PULMONARY AND SYSTEMIC CIRCULATION

(recall and comprehension)

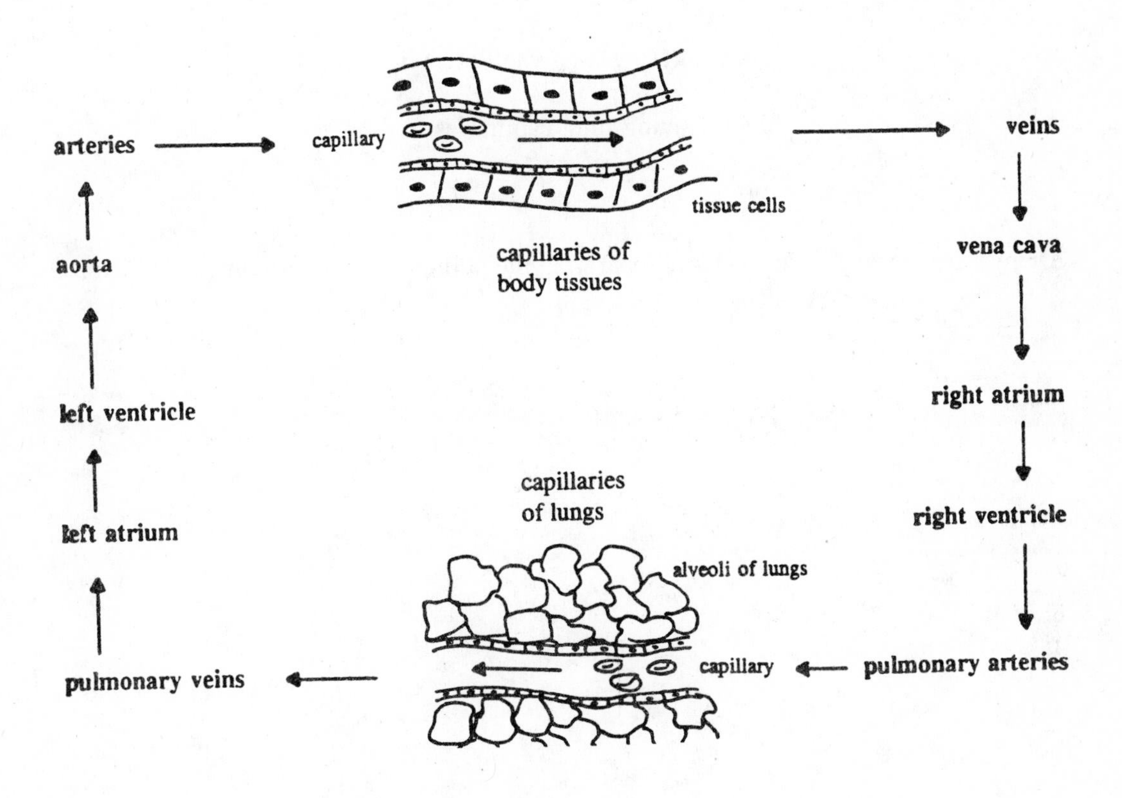

Figure 14-1. Pulmonary and Systemic Circulation

Insert the following arrows and labels into the diagram of pulmonary and systemic circulation in **Figure 14-1**:

1. **Arrows** showing the movement of O_2 and CO_2 in or out of capillaries in the body tissues.

2. **Arrows** showing the movement of O_2 and CO_2 in or out of capillaries in the air sacs of the lungs.

3. Is the blood **oxygenated (O) or deoxygenated (D)** at the following locations?

a. _____ aorta

b. _____ blood entering tissue capillaries

c. _____ blood leaving tissue capillaries

d. _____ vena cava

e. _____ blood entering lung capillaries

f. _____ blood leaving lung capillaries

4. **Draw a line** through **Figure 14-1**, separating the pulmonary and systemic circulations.

ACTIVITY 14-5: TRACING THE PATH OF BLOOD AROUND THE BODY

(recall)

Complete the following circular chart that follows a red blood cell from the thigh, to the heart, through the lungs, and back to the tissues of the thigh. **Name all major blood vessels, chambers of the heart, and heart valves** the blood cell passes on its journey. **Color code or label** the chart to show when the blood is **oxygenated** and **deoxygenated**.

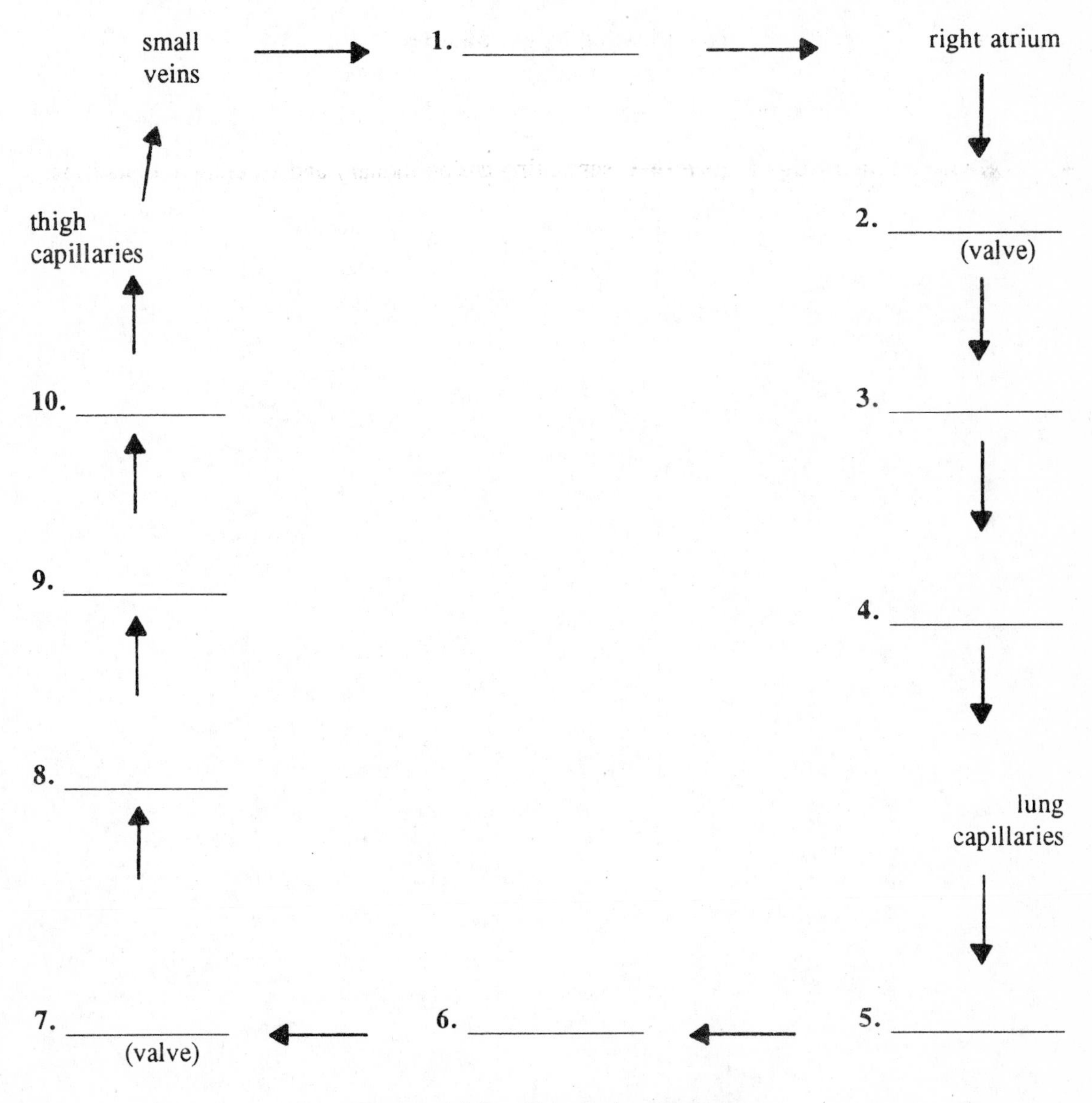

Figure 14-2. Circulation Pathway

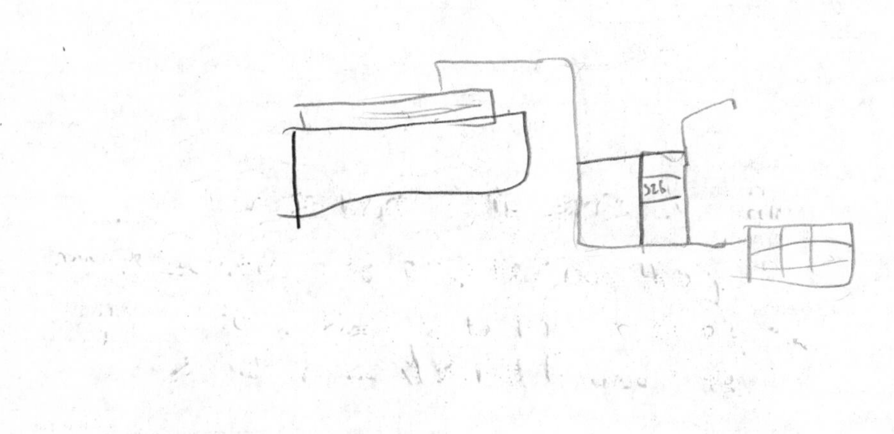

ACTIVITY 15-1: REFLEX ACTIONS

(comprehension)

You're walking barefoot along the boardwalk at your favorite seaside resort, gawking at members of the opposite sex, when, all of a sudden, you step on a sharp object. The following activities will be occurring in your nervous system. Place the following steps of a reflex action in the correct sequence from numbers one through seven.

Actions at the "whole body" level:

4 ____ muscles in your leg contract

1 ____ sensory nerve endings are stimulated

6 ____ brain senses pain

2 ____ sensory nerves transmit impulse to CNS

3 ____ motor nerves transmit impulses to leg muscles

3 ____ reflex action - foot moves

7 ____ members of opposite sex no longer seem so important

ACTIVITY 15-2: NERVE IMPULSE TRANSMISSION

(comprehension)

Remember when you were walking barefoot along the boardwalk at the seaside and you stepped on that sharp object? The following activities will be occurring in your nervous system. Place the following actions in the correct sequence from numbers one through nine.

Actions at the cellular level:

6 synaptic vesicles release neurotransmitter molecules

1 inside of sensory neuron membrane has a negative charge, outside is positive

2 sudden reversal in the electrical charge at the sensory neuron cell membrane

3 inside of sensory neuron membrane has a positive charge, outside is negative

8 sudden reversal in the electrical charge at the motor neuron

4 action potential travels along sensory neuron membrane

7 neurotransmitter molecules are received by the motor neuron

5 action potential received at spinal cord

9 muscle cell contracts

ACTIVITY 15-3: AT THE SYNAPSE

(comprehension)

In **YOUR OWN WORDS**, describe how a nerve impulse crosses a synapse. **Be complete.**

In your explanation, use the following terms: **neurotransmitter, vesicles, receptor proteins, synapse, calcium ions, and action potential**.

ACTIVITY 15-4: NERVOUS SYSTEM OVERVIEW

(recall)

Choose the most appropriate answer for the questions below. Answers can be used **only once**.

A. AUTONOMIC NERVOUS SYSTEM

B. CEREBELLUM

C. CORPUS CALLOSUM

D. GANGLION

E. HYPOTHALAMUS

F. PERIPHERAL NERVOUS SYSTEM

G. MEDULLA OBLONGATA

H. PINEAL BODY

I. NEUROTRANSMITTERS

J. WHITE MATTER

K. CEREBRAL HEMISPHERES

L. MOTOR CORTEX

M. OLFACTORY LOBES

N. MENINGES

O. THALAMUS

1. B integrates body position, motion, and balance
2. M interprets stimuli from receptors in the nasal epithelium
3. I acetylcholine is an example
4. G control center for breathing, heart rate, and blood pressure
5. J nerve fibers surrounded by a myelin sheath
6. D location of neuron cell bodies **outside** the central nervous system
7. A has sympathetic and parasympathetic divisions
8. C connects the right and left cerebral hemispheres
9. E controls homeostatic function, such as body temperature regulation, thirst, hunger, ion, and water balance
10. K center of consciousness and intelligence
11. H controls the timing of daily, season, and annual cycles
12. N protective covering of the brain and spinal cord
13. L issues commands to muscles

ACTIVITY 15-5: APPLYING YOUR KNOWLEDGE OF NERVE IMPULSE TRANSMISSION

(application)

1. You spray your house with an insecticide. Shortly afterwards, you observe roaches lying on the ground with their legs and wings twitching uncontrollably. What might the insecticide have done to the bugs' nervous systems to cause this reaction?

Insecticides inhabits the enzyme.

2. Multiple sclerosis is a disease in which nerve fibers in the central nervous system lose their myelin. Why would this loss be likely to affect the person's ability to control their skeletal muscles? Be specific.

ACTIVITY 16-1: SPECIFIC DEFENSES

(recall)

Complete this overview of the specific immune defenses by filling in each of the eight empty boxes with the correct **name of the cell, type of immunity, or function** involved.

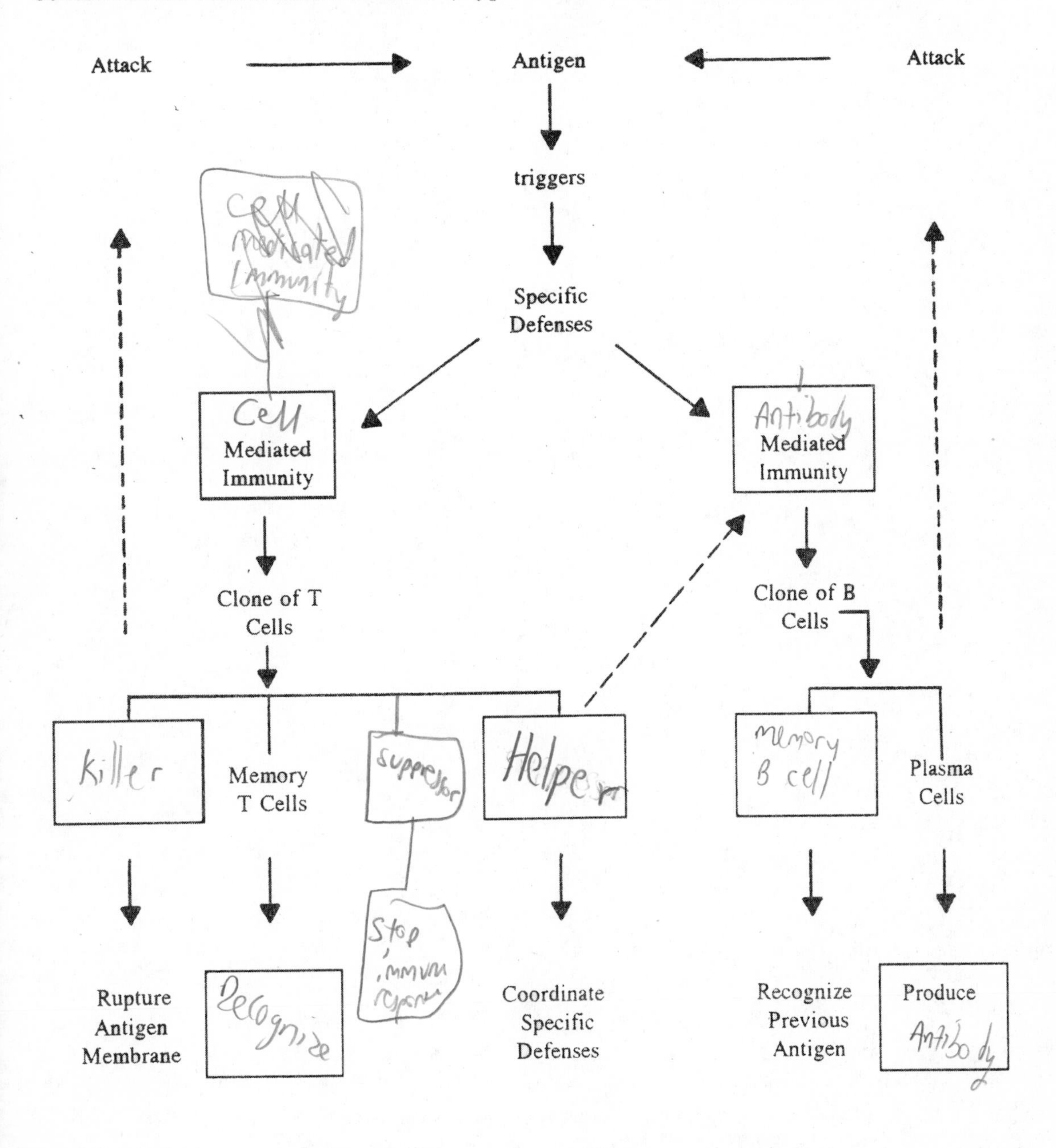

Figure 16-1. Overview of Specific Defenses

ACTIVITY 16-2: REVIEW OF THE IMMUNE RESPONSE

(recall and comprehension)

Rewrite these false statements to make them true statements. In your sentence, **you must change the UNDERLINED text!!**

1. An antibody will attack **any foreign invader** it passes in the circulatory system.

   specific invader

2. **Only B cells** can become memory cells.

   Both T and B

3. After a macrophage presents an antigen to the immune, the number of T cells in the body will **decrease**.

   increase

4. Pollen, dust, and animal fur are good examples of **antibodies**.

   antigens

5. After an immune response is no longer needed, many **helper T cells** remain in reserve.

   Memory T and B cells

6. **Mast cells** produce antibodies.

   plasma cells.

ACTIVITY 16-3: OVERVIEW OF THE IMMUNE RESPONSE

(recall)

Match the statements with the best answer below. Answers can be used **only once**.

A. LYMPHOCYTES
B. ANTIGEN
C. ANTIBODIES
D. PLASMA CELLS
E. MEMORY CELLS
F. MACROPHAGE
G. SKIN
H. THYMUS
I. HELPER T CELLS
J. MHC PROTEINS
K. SUPPRESSOR T CELLS
L. KILLER T CELLS
M. INFLAMMATION
N. NATURAL KILLER CELLS

1. _____ cell that coordinates the cell mediated and antibody mediated responses

2. _____ circulating cells still present many years after an infection is over

3. _____ large group of white blood cells that includes B cells, T cells, and others

4. _____ identifying markers on the membrane of body cells

5. _____ phagocytic cells that destroy invaders; part of the nonspecific defenses

6. _____ cells that destroy cancer cells as part of the nonspecific defenses

7. _____ part of the nonspecific defense mechanism that causes redness, increased body temperature, and swelling

8. _____ defensive proteins important in the chemical attack on antigens

9. _____ body's first line of defense

10. _____ foreign protein that can trigger an immune response

11. _____ cell that produces antibodies as part of the specific defense mechanism

12. _____ cell that stops the specific immune response when an infection is controlled

13. _____ organ in which T cells mature

14. _____ cell that attacks a specific invader with toxic chemicals

ACTIVITY 16-5: ANTIGENS AND ANTIBODIES OF THE A-B-O BLOOD SYSTEM

(comprehension)

Due to a clerical error, several samples of blood stored at the local blood bank may be incorrectly labeled as to blood type. The three tests listed below were conducted on each sample. Use your knowledge of antigen-antibody reactions to sort the samples into their correct blood types.

Test 1 - Unknown sample mixed with "anti-A" serum (contains "type A" antibodies)

Test 2 - Unknown sample mixed with "anti-B" serum (contains "type B" antibodies)

Test 3 - Unknown sample mixed with "anti-Rh" serum (contains antibodies against the Rh protein)

1. **Unknown Sample A:** When tested, agglutination (clumping) of red blood cells occurred in Tests 1, 2, and 3.

   Blood Type: ___________

2. **Unknown Sample B:** When tested, no agglutination occured in any of the three tests (1, 2, or 3).

   Blood Type: ___________

3. **Unknown Sample C:** When tested, agglutination occured in Tests 2 and 3, but not with Test 1.

   Blood Type: ___________

ACTIVITY 17-1: MALE AND FEMALE REPRODUCTIVE SYSTEMS

(recall)

Choose the most appropriate answer for the questions below. Answers may be used **more than once**.

A. URETHRA	**E. SEMINAL VESICLES**	**I. CERVIX**
B. UTERUS	**F. BULBOURETHRAL GLAND**	**J. PROSTATE GLAND**
C. EPIDIDYMIS	**G. SEMINIFEROUS TUBULES**	**K. VAS DEFERENS**
D. PENIS	**H. OVIDUCT (FALLOPIAN TUBE)**	**L. VAGINA**

1. _____ site of sperm formation
2. _____ sperm are stored here until they mature
3. _____ tube leaving the epididymis
4. _____ contributes fluid containing mucus, amino acids, and fructose
5. _____ contributes fluid with antibiotic enzymes
6. _____ carries urine or semen through the penis
7. _____ contributes a small amount of alkaline fluid just before ejaculation
8. _____ erection allows for insertion into vagina
9. _____ passageway for sperm into the female reproductive tract
10. _____ narrow opening into the uterus
11. _____ has a lining called the endometrium
12. _____ tube leading to the ovary; sperm meet the egg here
13. _____ organ in which the fetus develops

ACTIVITY 17-2: REVIEW OF THE HORMONES INVOLVED IN REPRODUCTION

(recall)

1. A pregnancy test is looking for this hormone. ____________________

2. This hormone increases sharply just before ovulation. ____________________

3. This hormone is produced by the corpus luteum and is present throughout pregnancy.

   ____

4. These hormones are not produced by the ovary or testes, but are found in both males and females. ____________________

5. This hormone is first present in males at puberty and is primarily responsible for beard growth. ____________________

6. This hormone is first present in females at puberty and is primarily responsible for the development of ovarian follicles. ____________________

7. This hormone is the "master controller" for the release of sex hormones.

   ____

8. This hormone stimulates breast development and determines distribution of subcutaneous fat layers in females. ____________________

ACTIVITY 17-3: HORMONAL CONTROL OF MALE AND FEMALE REPRODUCTION

(recall)

Fill in the blanks in the following table.

Hormone	Source of Production	Target in the Body	Effect
GnRH	hypothalamus	pituitary	production of FSH and LH
FSH (males)			
FSH (females)			1. 2.
LH (males)			
testosterone			
LH (females)			1. 2.
estrogen		1. 2.	1. 2.
progesterone			
HCG			

ACTIVITY 17-4: EVALUATING METHODS OF BIRTH CONTROL

(application, analysis, synthesis, and evaluation)

Effectiveness is one of the things to consider when choosing a birth control method. The numbers below compare the predicted (optimal) effectiveness of a birth control method with the typical effectiveness observed in the general population.

Method	Percent of Pregnancies per Year (Predicted Effectiveness)	Percent of Pregnancies per Year (Typical Effectiveness)
No birth control used	85 %	85%
Rhythm	9 %	20 %
Diaphragm and spermicide	6 %	18 %
Withdrawal	4 %	18 %
Male condom	2 %	12 %
IUD	1.5 %	3 %
Implants	.5 %	3 %
Birth control pills	.1 %	3 %
Sterilization	.2 %	.2 %

1. Using the graph paper in **Figure 17-1**, construct a **bar graph** that shows a **clear comparison** between the typical and predicted percentages of effectiveness for each method listed above.

   Place the percentages on the **Y axis** of the graph. Make sure to give the graph an appropriate title and label the X and Y axes.

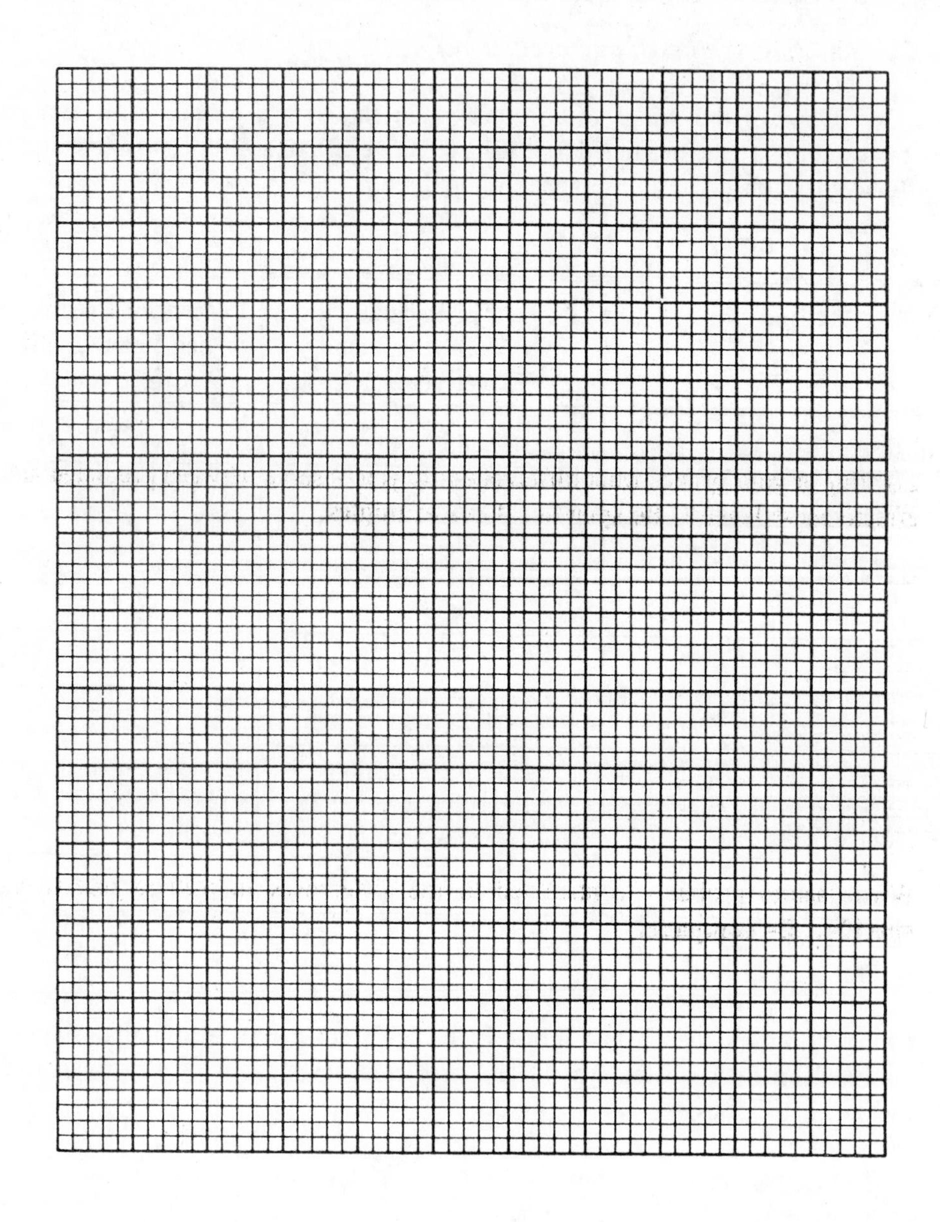

Figure 17-1.

2. Looking at your graph, would you say that **barrier methods** (such as diaphragms and condoms) are more or less effective than **chemical methods** (such as birth control pills)? **Explain your answer**.

3. Looking at your graph, what differences did you observe between predicted and typical effectiveness levels? **Be specific. Give examples.**

4. What factors probably contributed to the differences in method effectiveness? **Be specific. Be complete.**

ACTIVITY 17-5: HOW HORMONAL LEVELS AFFECT THE MENSTRUAL CYCLE

(comprehension)

Use information from **Figure 17-2** to answer the following questions concerning a typical human menstrual cycle.

1. On which day of the menstrual cycle is production of **LH greatest?**

2. What event in the menstrual cycle occurs **immediately after** LH reaches its peak production?

3. Which hormone increases steadily during the first half of the menstrual cycle?

4. What happens to the endometrium (uterine lining) while the hormone mentioned in question #3 is increasing?

5. When levels of estrogen and progesterone decline, what happens to the endometrium?

6. During which half of the reproductive cycle is progesterone at its **highest** levels?

7. On which day of the menstrual cycle would an **ovarian follicle** reach its **largest** size?

Challenge Question!!

8. Change the graph to show the following changes that occur in the reproductive cycle during pregnancy:

   a. thickness of endometrium
   b. progesterone levels
   c. FSH and LH levels

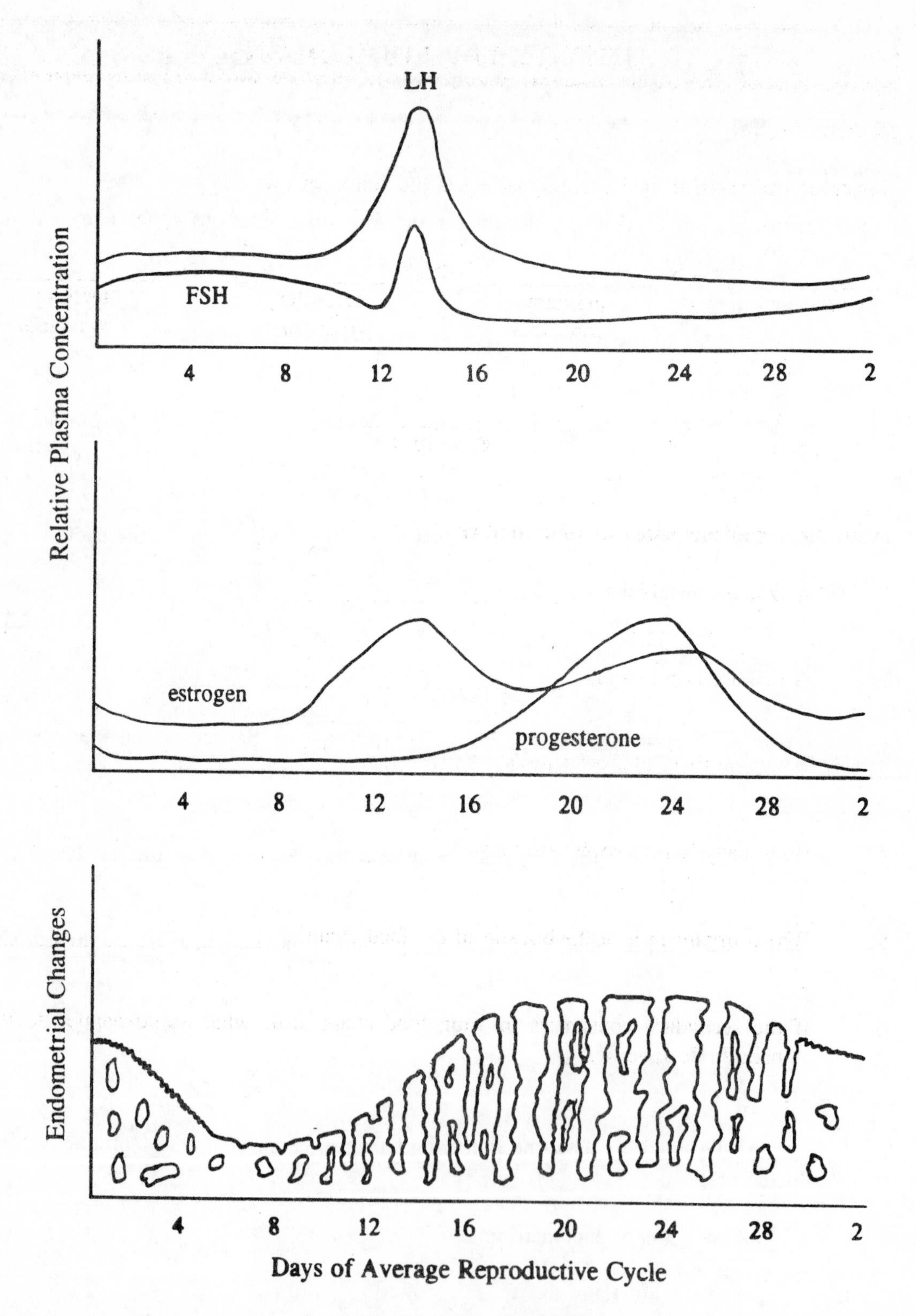

Figure 17-2. Hormone Production and Endometrial Changes During the Menstrual Cycle

ACTIVITY 18-1: FOOD CHAINS

(comprehension)

Construct a **terrestrial** food chain by filling in the blanks below:

____________	-->	____________	-->	____________	-->	____________
producer		**primary consumer**		**secondary consumer**		**tertiary consumer**

____

decomposer

From the organisms listed in your food chain:

1. Which are **herbivores**? ____________________

2. Which are **carnivores**? ____________________

3. Which perform **photosynthesis**? ____________________

4. How many **heterotrophs** are present? ________

5. Which organism is at the **bottom** of the food chain? ____________________

6. If the secondary consumers in **your** food chain died, what would happen to their remains? **Be specific**.

7. On your food chain, **circle** any organisms that could be food for a **decomposer**.

ACTIVITY 18-2: FOOD WEBS

(comprehension)

Examine the food web in **Figure 18-1** and answer the following questions:

1. Which organism(s) function as **both** primary and secondary consumers?

2. Which organism(s) function as **both** secondary and tertiary consumers?

3. The great blue heron is part of _______ different food chains.

4. How many **secondary consumers** are shown in this web? ______

5. If a largemouth bass dies of **natural causes**, it will become food for ______________.

6. How many different **producers** support this food web? ________

7. If all the great blue herons were removed from the ecosystem, the ____________ population would increase dramatically.

   This would, in turn, cause a **decrease** in the _______________ population.

   Would this affect the **trout and bluegill sunfish** populations? **Explain your answer**.

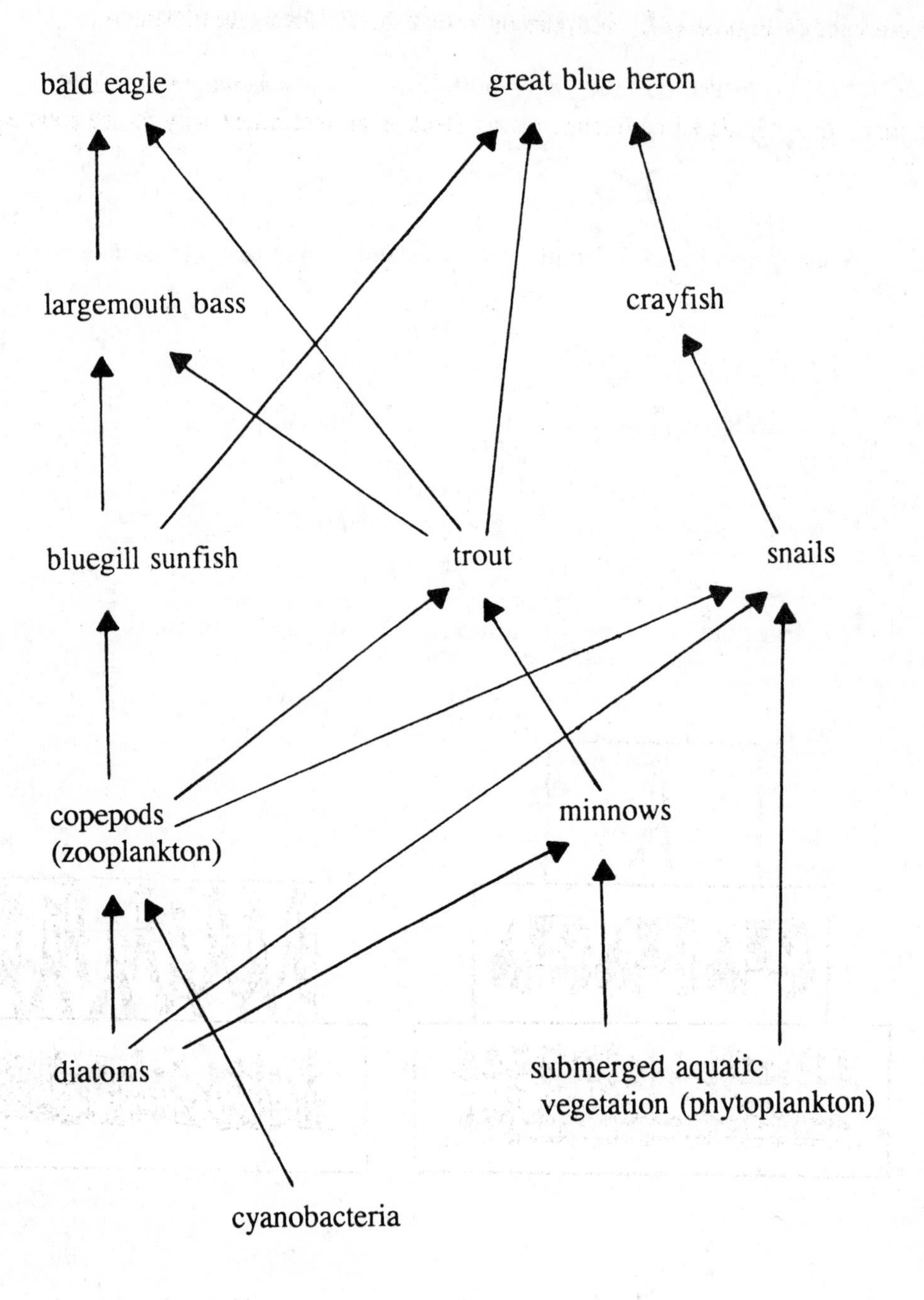

Figure 18-1. Simplified Wetlands Food Web

ACTIVITY 18-3: IS STEAK AN ECOLOGICAL LUXURY?

(comprehension)

In reference to **Figure 18-2, support or refute** the following statement:

"In many overpopulated countries, eating steak is an inefficient way to use food energy."

Figure 18-2. Energy Available in Long vs. Short Food Chains
From Human Biology: Condensed, by Bres and Weisshaar, p. 1-14. Copyright (©) 1997 (Education Resources). Reprinted by permission.

ACTIVITY 18-4: TROPHIC PYRAMIDS

(comprehension)

1. Arrange the following organisms into their appropriate levels on a trophic pyramid: **minnow, phytoplankton, striped bass, zooplankton,** and **bald eagles**.

2. What happens to the **energy** available as you move through the levels of your pyramid?

3. Draw an **asterisk (*)** next to the level of the pyramid that would show the **highest** amount of toxic chemical accumulation.

4. Give **three** examples of substances that might bioaccumulate in the ecosystem represented by **this trophic pyramid**.

5. Suppose the producer level of your pyramid contains **10,000 Kcal** of energy. How much of this energy will be available to feed your **secondary consumers**?

ACTIVITY 18-5: BIOACCUMULATION

(application)

Read the graph in **Figure 18-3** and answer the following questions:

1. How many parts per million of PCBs were found in the following organisms?

   zooplankton ________ ppm

   minnow ________ ppm

   lake trout ________ ppm

   herring gull ________ ppm

2. What **trend** does the graph show about the accumulation of PCBs in the food chain? **Explain your answer.**

   In your explanation, use the following terms: **toxic chemical, fat soluble, producer, consumer, and biological magnification.**

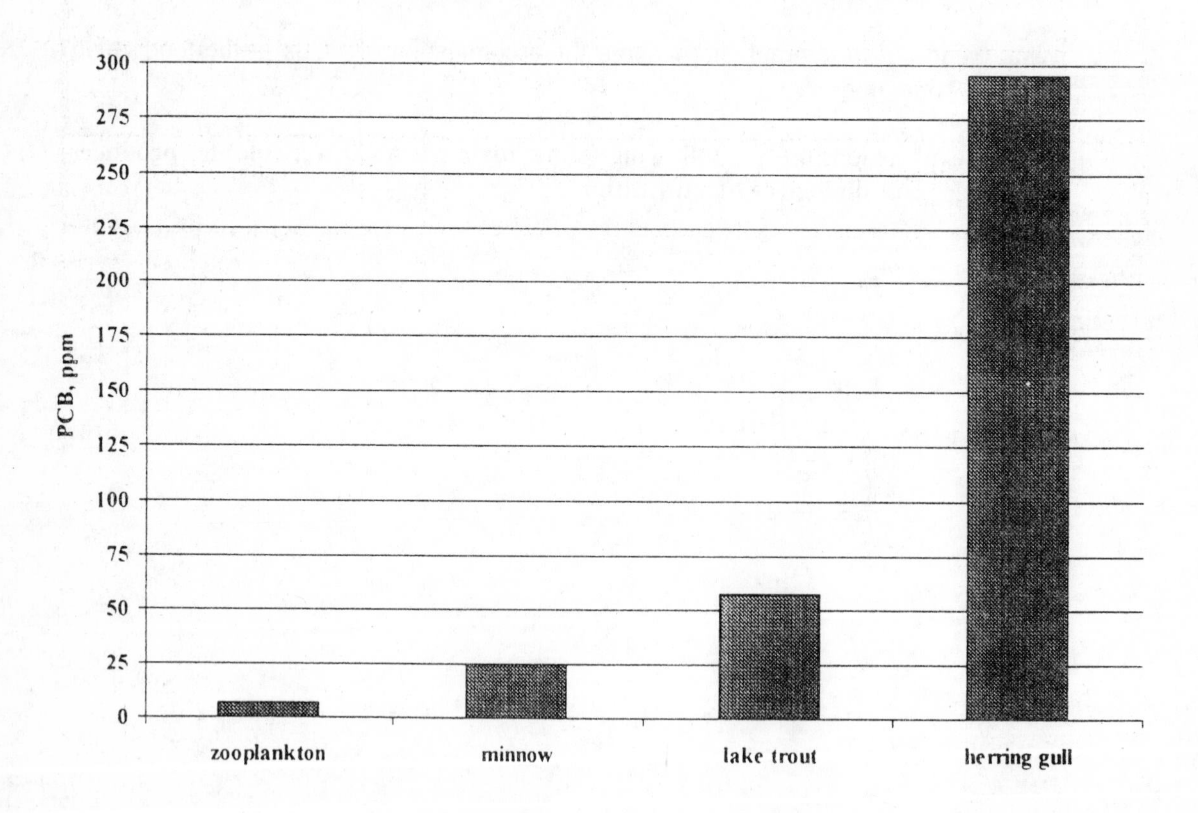

Figure 18-3. Bioaccumulation of PCBs

ACTIVITY 19-1: RENEWABLE AND NONRENEWABLE RESOURCES

(comprehension)

Classify each of the following as **renewable (R) or nonrenewable (N)**. **Explain each answer**.

	coal		wind		petroleum
	drinkable water		lumber		top soil
	solar energy		food crops		diamonds
	food animals		wave power to generate electricity		heating oil
	copper		aluminum		natural gas

ACTIVITY 19-2: DISTRIBUTION OF SOLID WASTE

(application, analysis, synthesis, and evaluation)

1. Using the information in **Table 19-1**, calculate the percent of America's total waste each item represents.

   To calculate percent of total waste:

$$\frac{\textbf{weight of item}}{\textbf{total weight}} \quad \textbf{X} \quad \textbf{100}$$

Table 19-1. Distribution of Solid Waste in the U.S.		
Item	**Weight (tons)**	**Percent of Total Waste**
Paper	73 million	
Yard Trimmings	35 million	
Aluminum	3 million	
Other Metals	14 million	
Glass	13 million	
Plastics	16 million	
Food	13 million	
Other (rubber, textiles, wood, leather, organic wastes, *etc.*)	28 million	
TOTALS		

2. On the graph paper in **Figure 19-1, plot the percent for each item.** You may use either a bar or line graph to represent the data.

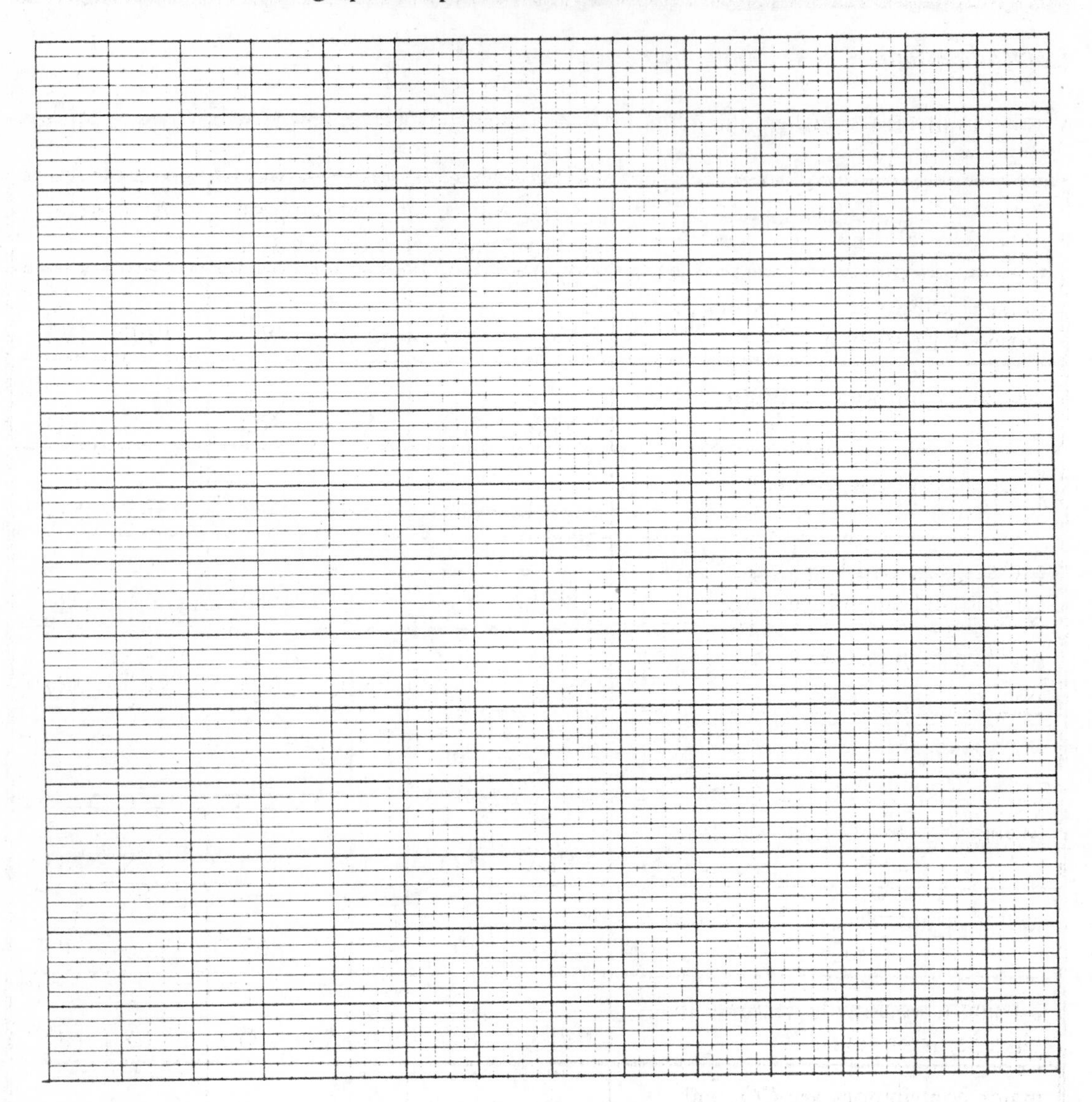

Figure 19-1. Distribution of Solid Waste in the U.S.

3. Choose **three items** from the list above. Explain how you can change your daily activities to decrease these components of solid waste.

ACTIVITY 19-3: GLOBAL ENVIRONMENTAL PROBLEMS

(comprehension)

Circle **yes or no** for each of the items. For each question, **circle one answer in each column**.

	Global Warming	**Thinning of Ozone Layer**	**Acid Rain**
causes earth to receive too much ultraviolet radiation	YES NO	YES NO	YES NO
caused by infrared radiation	YES NO	YES NO	YES NO
is an atmospheric problem	YES NO	YES NO	YES NO
causes decreased hatching success for fish and amphibian eggs	YES NO	YES NO	YES NO
problem is increased by using fossil fuels	YES NO	YES NO	YES NO
may cause flooding of coastal areas	YES NO	YES NO	YES NO
lowers the pH of rivers and lakes	YES NO	YES NO	YES NO
causes an increase in skin cancer	YES NO	YES NO	YES NO
causing a worldwide decrease in phytoplankton levels	YES NO	YES NO	YES NO
major contributors are CO_2 and methane	YES NO	YES NO	YES NO
CFCs are major contributors	YES NO	YES NO	YES NO
major contributors are nitrous oxides and sulfur dioxide	YES NO	YES NO	YES NO
may change global weather patterns	YES NO	YES NO	YES NO

ACTIVITY 19-4: MULTIPLE EFFECTS OF THE THINNING OZONE LAYER

(comprehension)

The thinning ozone layer causes health concerns (such as skin cancer) due to exposure to harmful ultraviolet (UV) radiation. UV light also damages the photosynthetic machinery of phytoplankton. Collapse of this producer population will cause the collapse of world-wide food webs, but would also cause a **DIFFERENT** but very serious problem for animal populations all over the world. This second problem is **NOT** related to energy transfer through food chains or food webs.

What is this **second** problem? **Explain your answer in detail.**

ACTIVITY 19-5: MODELING POPULATION GROWTH

(application, analysis, and synthesis)

On January 1, a population has one pair of RATS (**one male and one female**).

DATE	**POPULATION (## of rats)**
January 1	2
January 22	14

(The first litter is born to our happy family today. Each new generation will be born at three week intervals. Oh joy! Each new litter will have six males and six females.)

February 12	26
March 5	38
March 26	50
April 16	62

(The babies born on Jan 22 are now old enough to mate and have their own litters.)

May 7	146

(The six females born on Jan 22 gave birth to their first litters today. From now on, a new group of six females will mature and start breeding every three weeks.)

May 28	302
June 18	530
July 9	830
July 30	1202
August 20	2078
September 10	3890
October 1	7070
October 22	12,050
November 12	19,262
December 3	31,730
December 24	55,070

(In just three more weeks, the population will reach almost 100,000 rats!)

1. On the graph paper in **Figure 19-2**, **plot a graph** showing the growth of our rat population **from January 1 through December 24**.

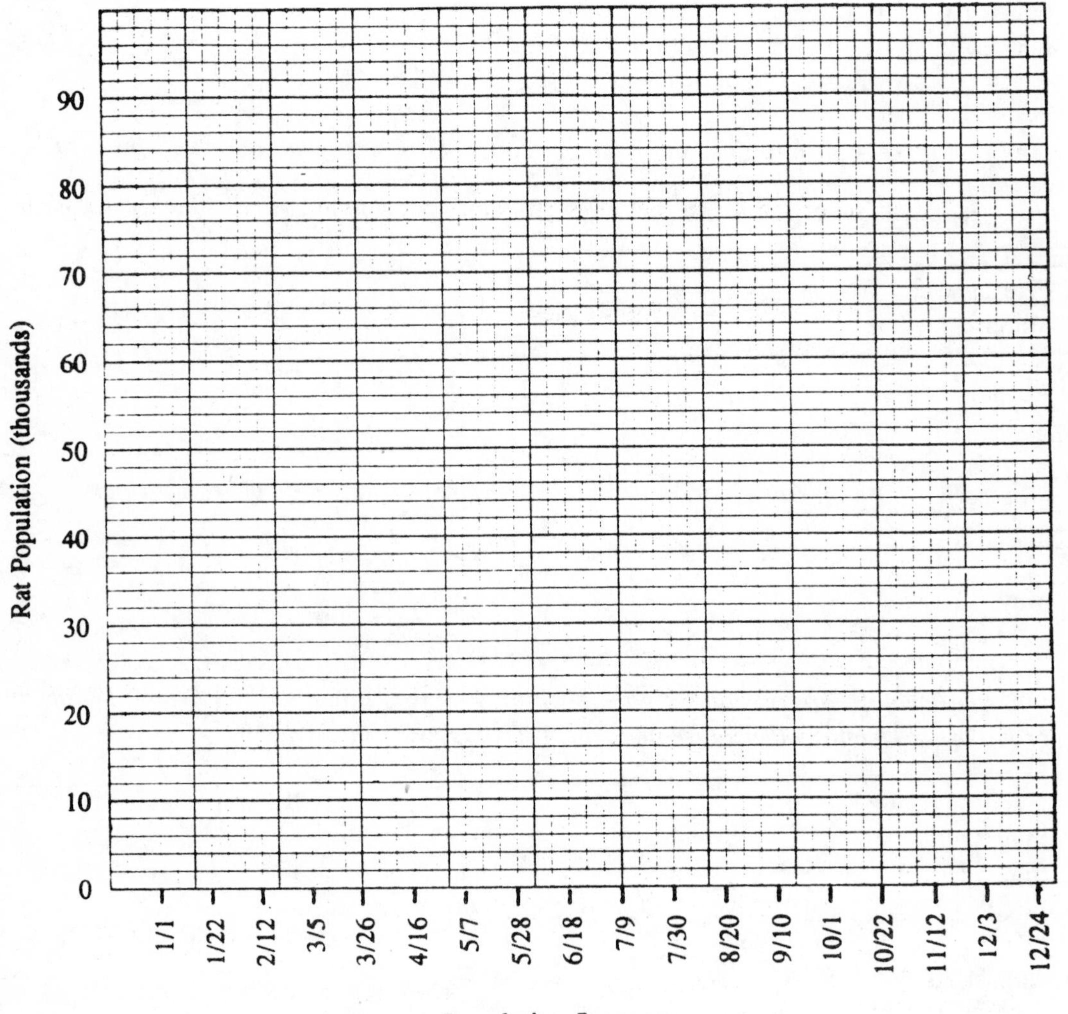

Figure 19-2. Rat Population Growth Curve

2. What can we learn about the growth of human populations by studying other animals?

3. Which soap opera title best describes the population model above? **Explain your answer**.

a. The Young and The Restless
b. Another World
c. All My Children
d. The Days of Our Lives

4. To determine when this population of rats will exceed its carrying capacity, you **need more information**. **List *all* the types of additional information** you would have to collect to determine the carrying capacity of our rat population.

ACTIVITY 19-6: PREDICTING POPULATION GROWTH FOR TWO COUNTRIES OF SIMILAR POPULATION SIZE

(application, analysis, synthesis, and evaluation)

1. Solve a problem in population growth comparing two countries with a similar population size. **Calculate the following information and enter it in Table 19-2.**

    a. Growth rate (GR) for each country
    b. Percent growth rate (PGR) for each country
    c. Doubling time (DT) for each country, rounded off to the nearest year

    Perform your calculations as shown below:

    a. **Growth rate (GR)**

    Growth rate (GR) = birth rate (BR) - death rate (DR)

    GR = ______ **per 1000 people in the population**

    b. **Percent growth rate (PGR)**

    $$PGR = \frac{GR}{1000} \times 100 = \_\_\_\_\_\_ \%$$

    c. **Doubling time (DT)**

    $$DT = \frac{70 \text{ (a population constant)}}{PGR} = \_\_\_\_\_\_ \text{ years}$$

Table 19-2. Population Comparison

	Kenya	Canada
2005 POPULATION (millions)	34	33
BR/1000	35	11
DR/1000	15	7
GR		
PGR		
DT		

2. Using the information from **Table 19-2**, **calculate the projected population** of each country and enter the information in **Table 19-3**. Stop your calculations when **Kenya** has doubled its population **FIVE times** and **Canada** has doubled **THREE times**.

Table 19-3. Predicted Populations of Kenya and Canada

Year	Kenya Projected Population (millions)	Year	Canada Projected Population (millions)
2005	34	2005	33

3. On the graph paper in **Figure 19-3**, plot a **graph** that shows your projected population figures. **Plan the scales** for the X and Y axes so that **both countries can be plotted on the same graph**. Make sure to include a **title** for your graph and to **label** the X and Y axes appropriately.

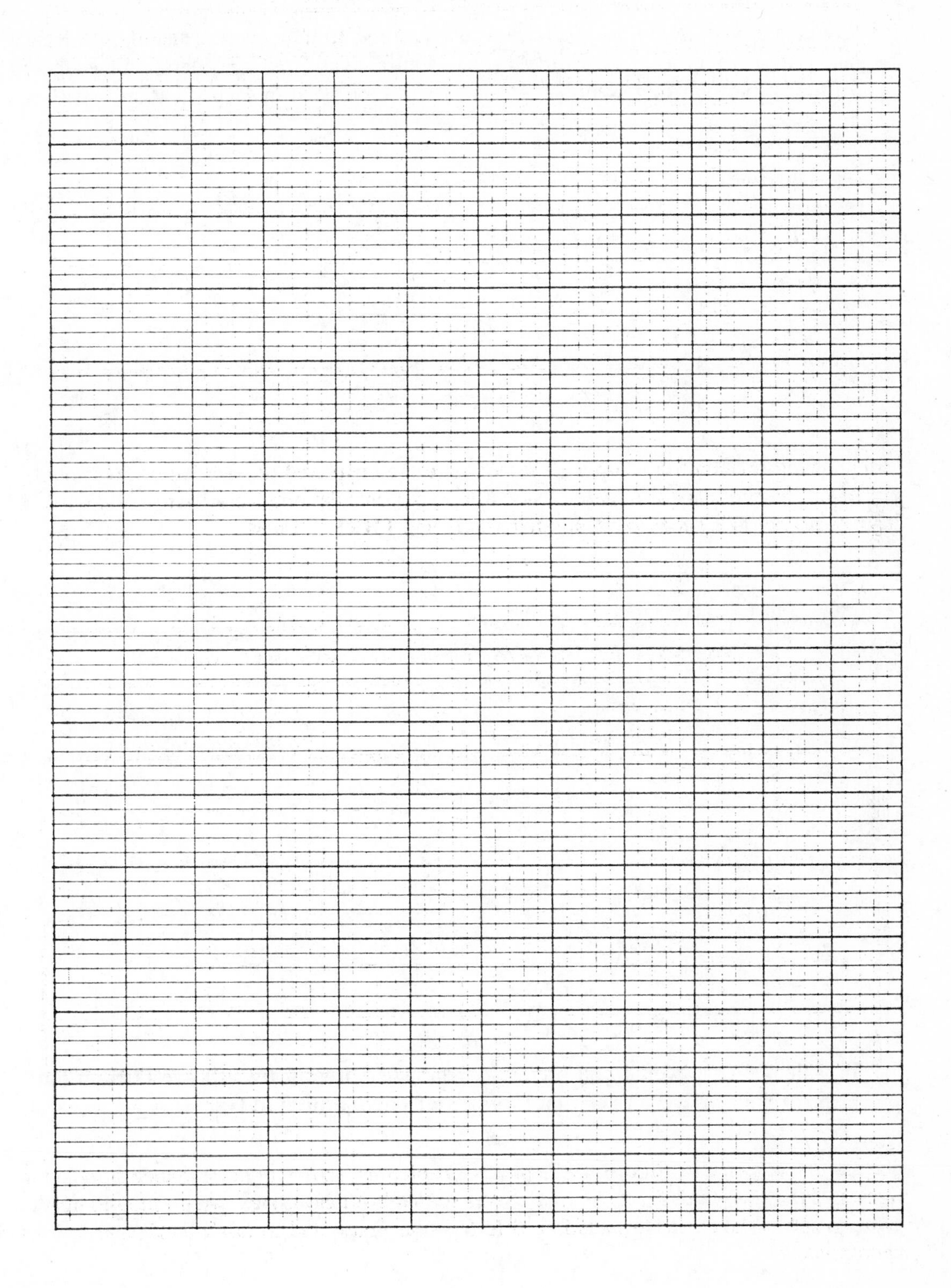

Figure 19-3. Comparison of the Projected Populations of Kenya and Canada

4. Looking at your graph, what advice would you give to the governments of **both** Kenya and Canada in terms of providing for their future. Take into consideration the amount of **food, fuel, housing, schools**, *etc.*, that will be needed in each situation.

Advice to Kenya:

Advice to Canada:

5. In which country would you prefer your grandchildren to live 100 years from now? **Explain** your answer **in relation to issues of population and resources**.

ACTIVITY 19-7: INTERPRETING POPULATION PROFILES

(comprehension) You are a specialist on population growth working for the United Nations. You are giving a speech on the world population problem. Using the information in the **population profiles** in **Figure 19-4**, **explain** to your audience whether **Country A or Country B** has the best chance of achieving zero population growth in the next **ten years**.

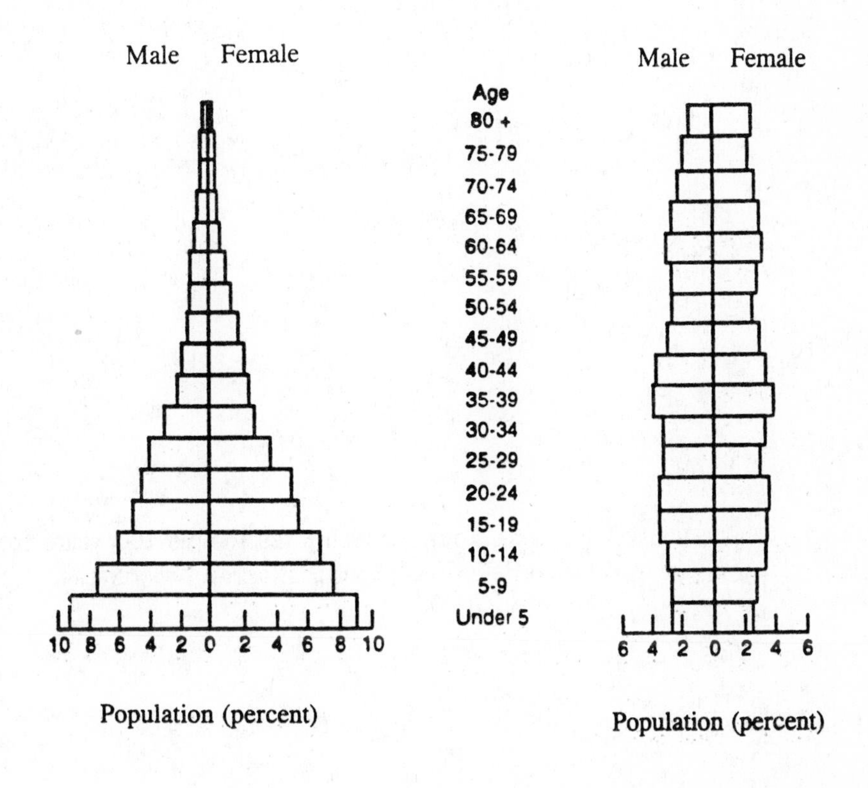

Figure 19-4. Comparing Population Profiles